WITHDRAWN

Innovation: A Very Short Introduction

'Despite the differences in surname, Mark Dodgson and I are brothers. I have known him and his faults all his life. How he wrote a book like this with David Gann I have no idea, but here it is, and a very good book too.

It tells a fascinating story, and one of growing importance. The ability to innovate is both expected and valued in the worlds of science and the arts: here we read about its importance in the field of business, and about how vastly our lives have changed – and continue to change – because of the innovation talents of individuals, and the innovation strategies of forward-thinking companies. There is a great deal here to fascinate not only those who are professionally engaged in business, but everyone who takes an intelligent interest in how the world is managed.'

Philip Pullman

Very Short Introductions available now:

For more information visit our web site:
www.oup.co.uk/general/vsi/

Mark Dodgson and David Gann

INNOVATION

A Very Short Introduction

1460-710

JAN 2 0 2012

OXFORD
UNIVERSITY PRESS

Great Clarendon Street, Oxford OX2 6DP

Oxford University Press is a department of the University of Oxford.
It furthers the University's objective of excellence in research, scholarship,
and education by publishing worldwide in

Oxford New York

Auckland Cape Town Dar es Salaam Hong Kong Karachi
Kuala Lumpur Madrid Melbourne Mexico City Nairobi
New Delhi Shanghai Taipei Toronto

With offices in

Argentina Austria Brazil Chile Czech Republic France Greece
Guatemala Hungary Italy Japan Poland Portugal Singapore
South Korea Switzerland Thailand Turkey Ukraine Vietnam

Oxford is a registered trade mark of Oxford University Press
in the UK and in certain other countries

Published in the United States
by Oxford University Press Inc., New York

British Library Cataloguing in Publication Data

Data available

Library of Congress Cataloging in Publication Data

Data available

Typeset by SPI Publisher Services, Pondicherry, India
Printed in Great Britain by
Ashford Colour Press Ltd, Gosport, Hampshire

ISBN 978-0-19-956890-1

3 5 7 9 10 8 6 4

For Yo and Anne

Contents

Preface

When we were born, not so very long ago, there were no
information technologies or television companies, and airline
travel was rare and luxurious. Our parents were born into a world
even more different than today's, where television had yet to be
invented, and there was no penicillin or frozen food. When our
grandparents were born, there were no internal combustion
engines, aeroplanes, cinemas, or radios. Our great grandparents
lived in a world with no light bulbs, cars, telephones, bicycles,
refrigerators, or typewriters, and their lives probably had more in
common with a Roman peasant than with ours. In the relatively
short period of 150 years, our lives at home and work have been
completely transformed by new products and services. The reason
why the world has changed so much can be explained in large part
by innovation.

This *Very Short Introduction* defines innovation as ideas,
successfully applied, and explains why it has the ability to affect us
so profoundly. It will describe how innovation occurs, what and
who stimulates it, how it is pursued and organized, and what its
outcomes are, both positive and negative. It will argue that
innovation is essential to social and economic progress, and yet
that it is hugely challenging and besotted with failure. It describes
how innovation has many contributors and takes different forms,
adding to its complexity. It provides an analysis of the innovation

process; the ways organizations marshal their resources to innovate, and the eventual outcomes of innovation, which can take a number of forms.

Innovations are found not only in the activities organizations do, but how they do them. The innovation process is presently going through a period of change, stimulated in large part by the opportunities of using new internet and visualization technologies to access ideas distributed from around the world. The potential sources of innovation are growing rapidly. There are, for example, more scientists and engineers alive today than in past history combined. Furthermore, the locus of innovation is changing as economies become dominated by service sectors and the ownership of, or access to, knowledge is ever more valuable compared to physical assets. Innovation is becoming more internationalized, with important new sources emerging in China, India, and elsewhere outside of the industrial powers of Europe, North America, and Japan. We explore the extent to which our understanding of innovation, developed over the past century or more, might be applied to deal with the restless transformations and turbulence we will witness in the global economy in the future.

The first three chapters explain what innovation is, its importance, and its outcomes. The subsequent chapters examine the contributors to innovation and how it is organized, and speculate on its future.

Our understanding of innovation is based on our research into countless innovative organizations around the world and our learning from the accumulated efforts of numerous scholars in the international innovation research community. Our grateful thanks are extended to all those innovators, and students of innovation, who make our journey so exciting and rewarding. We especially acknowledge Irving Wladawsky-Berger and Gerard Fairtlough, two great innovators who have had profound influences on our thinking.

List of illustrations

Chapter 1
Josiah Wedgwood: the world's greatest innovator

We begin with a study of an exemplary innovator, a person who tells us a great deal about the innovator's agenda. He established an enduring, high-profile company creating substantial innovations in the products made, the ways they were produced, and the manner in which they created value for himself and his customers. He made significant contributions to building national infrastructure, helped create a dynamic regional industry, pioneered new export markets, and positively influenced government policies. His outstanding scientific contribution was recognized by election as a Fellow of the Royal Society. He was a marketing genius, and bridged the scientific and artistic communities by a wholly new approach to industrial design. His most important contribution lay in the way he improved the quality of life and work in the society in which he lived. He is the potter Josiah Wedgwood (1730–95).

Born in modest circumstances into a family of Staffordshire potters, Wedgwood was the youngest of 13 children, and his father died when he was young. He was put to work as a potter when he was 11. He suffered badly from smallpox as a child and this had a big impact on his life. As William Gladstone put it, his disease 'sent his mind inwards, it drove him to meditate upon the laws and secrets of his art ... and made for him ... an oracle of his own inquiring, searching, meditative, fruitful mind'. For the first part

1. The world's greatest innovator

of his career, he worked in a number of partnerships, studying every branch of the manufacture and sale of pottery. By the time Wedgwood began his own business, aged 29, he had mastered every aspect of the pottery industry.

In his mid-30s, the lameness resulting from smallpox proved too much of a constraint, so he had his leg amputated, without, of course, the aid of antiseptic or anaesthetic. As testament to his energy and drive, he was writing letters within a couple of days. A few weeks later, he suffered the tragic loss of one of his children, but he was back at work within a month of the operation.

By the mid-18th century, the European ceramics industry had been dominated by Chinese imports for around 200 years. Chinese porcelain, invented nearly a thousand years before, achieved a quality in material and glaze that could not be matched. It was much prized by the wealthy, but was too expensive for the expanding industrial classes whose incomes and aspirations were growing during this period of the Industrial Revolution. Trade restrictions on Chinese manufactures further increased the price of imports into Britain. The situation was ripe for innovation to provide attractive, affordable ceramics for a mass market.

Wedgwood was a product innovator, constantly searching for innovation in the materials he used, and in glazes, colours, and design forms of his wares. He applied extensive trial-and-error experiments to continually improve quality by removing impurities and making results more predictable. His favourite motto was 'Everything yields to experiment'. Some innovations resulted from incremental improvements to existing products. He refined a new cream-coloured earthenware being developed in the industry at the time, transforming it into a high-quality ceramic that was very versatile in that it could be thrown on a wheel, turned on a lathe, or cast. After producing a dinner service for Queen Charlotte, wife of George III, and receiving her approval, he named this innovation 'Queen's Ware'. Other innovations were more radical. In 1775, after around 5,000 recorded experiments that were often difficult and expensive, he produced Jasper, a fine ceramic, commonly blue in colour. This was one of the most significant innovations since the invention of porcelain. His major product innovations were still being produced by the Wedgwood company more than 200 years later.

He collaborated with numerous artists and architects in the design of his products, including George Hepplewhite, the furniture maker; Robert Adam, the architect; and George Stubbs, the artist. One of his great achievements was the application of design to the everyday. The renowned sculptor John Flaxman, for example, produced inkstands, candlesticks, seals, cups, and teapots. Products that were previously unattractive were made elegant.

Wedgwood searched everywhere for ideas for designs, from customers, friends, and rivals. He looked in museums and great houses, and trawled antique shops. One valuable source of designs was a coterie of amateur artists amongst well-bred women. Part of Wedgwood's successful approach to working with artists, according to Llewellyn Jewitt, his 19th-century biographer, lay in his effort 'to sharpen the fancy and skill of the artist by a collision with the talents of others'.

In a speech by William Gladstone, a generation after Wedgwood's death, he says of the potter:

> His most signal and characteristic merit lay ... in the firmness and fullness of his perception of the true law of what we term industrial art, or in other words, the application of the higher art to industry: the law which teaches us to aim first at giving to every object the greatest possible degree of fitness and convenience for its purpose, and next making it the vehicle for the highest degree of beauty, which compatibility with the fitness and convenience it will bear: which does not substitute the secondary for the primary end, but recognises as part of the business the study to harmonize the two.

In his manufacturing process innovations, Wedgwood introduced steam power into his factory, and as a result the Staffordshire pottery industry was the earliest adopter of this new technology. Steam power brought many changes to production processes. Previously the potteries were distant from the mills that provided power for mixing and grinding raw materials. Having power on-site significantly reduced transportation costs. It also mechanized the processes of throwing and turning pots, previously driven by foot or hand wheels. Technology enhanced efficiency in the way the use of lathes to trim, flute, and checker products improved production throughput.

He was preoccupied with quality, and spent vast amounts on pulling down and rebuilding kilns to improve their performance. Famously intolerant of poor product quality, legend has him prowling the factory smashing substandard pots and writing in chalk 'this won't do for Josiah Wedgwood' on offending workbenches.

One of the perennial challenges of making ceramics was measuring high temperatures in kilns in order to control the production process. Wedgwood invented a pyrometer, or thermometer, that recorded these temperatures, and for this achievement he was elected a Fellow of the Royal Society in 1783.

Many of Wedgwood's most popular products were produced in large numbers in plain shapes, which were then embellished by designers to reflect current trends. Other, more specialist products were produced in short, highly varied batches, quickly changing colour, fashion, style, and price as the market dictated. He subcontracted the manufacture of some products and their engraving to reduce his own inventory. When orders exceeded his production capacity, he outsourced from other potters. Wedgwood's innovative production system aimed to minimize proprietary risk and reduce fixed costs. He was highly aware of costs, having at one time complained that his sales were at an all-time high, yet profits were minimal. He studied cost structures and came to value economies of scale, trying to avoid producing one-off vases 'at least till we are got into a more methodicall way of making the same sorts over again'.

Wedgwood was an innovator in the way work was organized. His organizational innovations were introduced into an essentially peasant industry, with primitive work practices. When Wedgwood founded his main Staffordshire factory, Etruria, he applied the principles of the division of labour espoused by his contemporary, Adam Smith. Replacing previous craft production techniques, where one worker produced entire products, specialists concentrated on one specific element of the production process to enhance efficiency. Craftsmanship improved, allowing artists, for example, to improve the quality of designs, and innovation flourished. One of his proudest boasts was that he had 'made artists of mere men'.

Wedgwood paid slightly higher wages than the local average and invested extensively in training and skills development. In return, he demanded punctuality, introducing a bell to summon workers and a primitive clocking-in system, fixed hours, and constant attendance; high standards of care and cleanliness; avoidance of waste; and a ban on drinking. Wedgwood was conscious about health and safety, especially in relation to the ever-present dangers of lead poisoning. He insisted on proper cleaning methods, work attire, and washing facilities.

As a business innovator, Wedgwood created value by engaging with external parties in a number of ways. He innovated in sources of supply and distribution, astutely used personal and business partnerships to advantage, and introduced a remarkable number of marketing and retailing innovations.

Wedgwood sought the best-quality raw materials from wherever he could find them. In what today would be called 'global sourcing', he purchased clay from America in a deal struck with the Cherokee nation, from China, and the new colony in Australia.

He had a wide range of friends with very diverse interests upon whom he drew in his business dealings. Wedgwood belonged to a group of similarly minded polymaths who became known as the Lunar Men, because of their meeting during the full moon. Along with Wedgwood, they comprised a core of Erasmus Darwin, Matthew Boulton, James Watt, and Joseph Priestley. The friendship and business partnership with Boulton was particularly influential on Wedgwood's thinking about work organization, as he observed the efficiency, productivity, and profitability of the Boulton and Watt factory making steam engines in Birmingham. Jenny Uglow's book on the Lunar Men argues that they were at the leading edge of almost every movement of their time, in science, in industry, and in the arts. She evocatively suggests that: 'In the time of the Lunar men, science and art were not separated, you could be an inventor and designer, an experimenter and a poet, a dreamer and an entrepreneur all at once.'

Although Wedgwood had somewhat contradictory views on the ownership of intellectual property, he encouraged collaborative research and was a proponent of what today would be called 'open innovation'. In 1775, he proposed a cooperative programme with fellow Staffordshire potters to solve a common technical problem. It was a plan for what was the world's first collaborative industrial research project. The scheme failed to get off the ground, but it illustrates a desire to use a form of organization that was not again explored for over a century.

Wedgwood was the first in his industry to mark his name on his wares, denoting ownership of the design, but he disliked patents, and only ever owned one. Speaking of himself, he explains his approach:

> When Mr. Wedgwood discovered the art of making Queen's ware ... he did not ask for a patent for this important discovery. A patent would greatly have limited its public utility. Instead of one hundred manufactures of Queen's ware there would have been one; and instead of an exportation to all quarters of the world, a few

pretty things would have been made for the amusement of the people of fashion in England.

The period of the Industrial Revolution was one of great optimism as well as social upheaval. Consumption and lifestyle patterns changed as industrial wages were paid and new businesses created novel sources of wealth. The population of England doubled from around 5 million in 1700 to 10 million in 1800. Until the 18th century, English pottery had been functional; mainly crude vessels for storing and carrying. Pots were crudely made, ornamented in an elementary way, and glazed imperfectly. The size and sophistication of the market developed throughout the 18th century. Stylish table accessories were in huge demand in the burgeoning industrial cities and increasingly wealthy colonies. Drinking tea, and more fashionable coffee and hot chocolate, joined the traditional British pastime of imbibing beer as a national characteristic.

Wedgwood sought to meet and shape this burgeoning demand in a number of ways. Initially he sold his completed wares to merchants for resale, but he opened a warehouse in London, followed by a showroom that took direct orders. Browsing customers commented on the wares on display, and Wedgwood took particular note of criticisms of uneven quality, explaining his devotion to researching how to achieve better consistency. The showroom, run by Wedgwood's close friend, Thomas Bentley, became a place for the fashionable to be seen, and major new collections were visited by royalty and aristocracy. Bentley expertly interpreted new trends and tastes, informing design and production plans back in Staffordshire.

Wedgwood assiduously sought patronage from politicians and aristocrats: what he called his 'lines, channels, and connections'. He produced a 952-piece dinner service for Catherine the Great, Empress of Russia, shamelessly using her patronage in his advertising. His view was that if the great and the good bought his products, the new middle classes, merchants and professionals, and even some of the wealthier lower classes, artisans and tradespeople, would aspire to emulate them.

An astonishing number of retailing innovations were introduced by Wedgwood and Bentley, including the display of wares set out in full dinner service, self-service, catalogues, pattern books, free carriage of goods, money-back guarantees, travelling salesmen, and regular sales, all aiming 'to amuse, and divert, and please, and astonish, nay, and even to ravish the Ladies'. Jane Austen wrote of the pleasure of the safe delivery of a Wedgwood order.

Wedgwood's international marketing efforts were pioneering. When he started his business, it was rare for Staffordshire pottery to reach London, let alone overseas. To sell in international markets, he again used the strategy of courting royalty by using his English aristocratic connections as ambassadors. By the mid-1780s, 80% of his total production was exported.

Products were not sold on the basis of low prices. Wedgwood's products could be two or three times as expensive as his competitors'. As he put it, 'it has always been my aim to improve the quality of the articles of my manufacture, rather than to lower their prices'. He was contemptuous of price cutting in the pottery industry, writing to Bentley in 1771:

> the General Trade seems to me to be going to ruin on the gallop . . . low prices must beget a low quality in their manufacture, which will beget contempt, which will beget neglect, and disuse, and there is an end of the trade.

Wedgwood's innovations extended into many other areas. He expended substantial efforts in building the infrastructure supporting the manufacture and distribution of his products and those in his industry. He devoted significant amounts of time and money to improving communications and transportation, especially with the ports that supplied raw materials and provided his routes to market. He promoted the development of turnpike roads and became centrally engaged in the construction of major canals. He actively lobbied the government on trade and industry policy and helped form the first British Chamber of Manufacturers.

Wedgwood's legacy extended well beyond his company. He had an enormous impact on the Staffordshire Potteries more generally, in what today might be called an innovative 'industrial cluster'. Pottery production in Staffordshire developed rapidly due to the efforts of numbers of firms, such as Spode and Turner, but Wedgwood was the acknowledged leader of the industry.

His 19th-century biographer, Samuel Smiles, wrote of the change from the 'poor and mean villages' brought about by Wedgwood's innovations:

> From a half-savage, thinly peopled district of some 7000 persons in 1760, partially employed and ill remunerated, we find them increased, in the course of some twenty-five years, to about treble the population, abundantly employed, prosperous, and comfortable.

Wedgwood's contributions to public life included improving the education, health, diet, and housing of his employees. Etruria's 76 homes were, in their time, considered a model village.

Wedgwood built a dynasty. He inherited £20 from his father, and when he died he left one of the finest industrial concerns in England with a personal worth of £500,000 (around £50 million at present prices). Wedgwood's children used their good fortune well. One son created the Royal Horticultural Society and another contributed centrally to the development of photography. Wedgwood's wealth was used to great effect to fund the studies of his grandson, Charles Darwin.

The Wedgwood case raises a number of core issues that we shall be examining in this Very Short Introduction and reveals the approach to innovation we shall be taking. We focus on the organization, the mechanism for creating and delivering innovation. The individuals and their personal connections whose importance is so clearly shown in the Wedgwood case will be discussed here only to the extent to which they contribute to organizational outcomes. We do not discuss the meanings of innovation for us individually. Nor do we adopt the perspective of the user of innovation, although we shall argue that innovative organizations need to try to understand how innovations are consumed and for what purpose. With this observation in mind, Wedgwood shows us that innovation occurs in many forms and ways. It occurs in what organizations produce: their products and services. It is found in the ways in which organizations produce: in their production processes and systems, work structures and practices, supply arrangements, collaboration with partners, and very importantly how they engage with and reach customers. Innovation also happens in the context within which organizations operate: for example, in regional networks, supportive infrastructure, and government policies.

Wedgwood illustrates an enduring truth about innovation: it involves new combinations of ideas, knowledge, skills, and resources. He was a master at combining the dramatic scientific, technological, and artistic advances of his age with rapidly changing consumer demand. Gladstone said: 'He was the greatest man who ever, in any age or in any country, applied himself to the important work of mixing art with industry.' The way Wedgwood merged technological and market opportunities, art and manufacturing, creativity and commerce, is, perhaps, his most profound lesson for us.

Chapter 2
Joseph Schumpeter's gales of creative destruction

All economic and social progress ultimately depends on new ideas that contest the introspection and inertia of the status quo with possibilities for change and improvement. Innovation is what happens when new thinking is successfully introduced in and valued by organizations. It is the arena where the creation and application of new ideas are formally organized and managed. Innovation involves deliberate preparations, objectives, and planned benefits for new ideas that have to be realized and implemented in practice. It is the theatre where the excitement of experimentation and learning meets the organizational realities of limited budgets, established routines, disputed priorities, and constrained imagination.

There are a great many ways of understanding innovation that provide a wide range of rich insights and perspectives. The variety of different analytical lenses used depends on the particular innovation issues being studied. Some analyse the extent and nature of innovation: whether any change is incremental or radical, how it sustains or disrupts existing ways of doing things, and if it occurs in whole systems or their components. Other analyses are concerned with how the focus of innovation changes over time, that is, from the development of new products to their manufacture, their patterns of diffusion, how particular design configurations, such as video recorders and music players, come to dominate in the market, and how organizations appropriate value from innovation.

Defining

The English language allows broad definitions of innovation, which can be both helpful and confusing. This is helpful in as much as it can usefully cover a wide range of activities and is confusing for the same reason – the word can be used promiscuously. Even the relatively simple definition of innovation we use – ideas, successfully applied – raises questions. What is 'success'? Time is influential, and innovations may be initially successful and eventually fail, or vice versa. What does 'applied' imply? Is it applied within a single part of an organization, or diffused internationally amongst a massive group of users? What and who are the sources of 'ideas'? Can anyone lay claim to them, especially as they inevitably combine new and existing thinking?

Typologies of innovation also face difficulties because of blurred boundaries and overlaps between categories. Innovation occurs in products, for example in new cars or pharmaceuticals; and services, for example in new insurance policies or in health monitoring. But many service firms describe their offerings as products, such as new financial products. Innovation occurs in operational processes, in the way new products and services are delivered. These processes may take the form of equipment and machinery, which are the providers' products, and logistics in the form of transportation, which are providers' services.

There are some similar definitional problems when thinking about levels of innovation. A minor innovation for one organization may be substantial for another. It is difficult in practice to develop anything but a nominal scale of the differences between levels of innovation, and categorization is best thought of as ideal types along a continuum. Most innovations are incremental improvements – ideas used in new models of existing products and services, or adjustments to organizational processes. They might include the latest versions of particular software packages, or decisions to add more representatives from the marketing department to development teams. Radical innovations change the nature of products, services, and processes. Examples include the development of synthetic materials, such as nylon, and decisions to use open-source software to encourage community development of new services, rather than doing it proprietarily. At the highest level, there are rarer, periodic transformational innovations, which are revolutionary in their impact and affect the whole economy. Examples would be the development of oil as an energy source, or the computer or internet.

We think about innovation as ideas, successfully applied in organizational outcomes and processes. Innovation can be thought of as practical and functional: the outcomes of innovation are new products and services, or they are the organizational processes supporting innovation that occur in departments such as R(esearch)&D(evelopment), Engineering, Design, and Marketing. Innovation can also be thought of more conceptually: the outcomes of innovation are enhanced knowledge and judgement, or they are the processes that support the capacity of organizations to learn.

We have chosen to focus on innovations other than those described as 'continuous improvement' that tend to be routine and highly incremental in nature. Although these small improvements are cumulatively important, our concern lies rather with ideas that stretch and challenge organizations as they attempt to survive and thrive. By concentrating on innovations beyond the ordinary that occur in both the outcomes of organizational efforts and the processes that produce them, we capture a great degree of what is generally understood to be innovation.

Importance

The reason why innovation is so important has to be seen in the context of the relentless demands made of contemporary organizations as they face the challenges of a complex and turbulent world. Innovation is crucial for their continuing existence as they struggle to adapt and evolve to deal with constantly changing markets and technologies.

In the private sector, the threat of new competitors in globalized markets is ever present. In the public sector, the demand for efficiencies and enhanced performance is continual, as governments attempt to manage demands for expenditure to improve the quality of life that exceed their incomes. The motivation to innovate in all organizations is stimulated by the knowledge that if they are not capable of innovation, others are: new players that may threaten their very existence. Simply, if organizations are to progress – to develop and grow, become more profitable, efficient, and sustainable – they need to successfully implement new ideas. They have to be continually innovative. As the economist Joseph Schumpeter put it, at its most blunt, innovation 'offers the carrot of spectacular reward or the stick of destitution'.

One of the features of innovation is that it can be found in every organization. Although the cost of innovation can be very high – it can, for example, cost up to $800 million to develop a new pharmaceutical – new ideas can be successfully implemented cheaply. It is not only high-tech firms making semiconductors or working with biotechnology that rely on innovation in their businesses, it is all parts of the economy. Insurance firms and banks continually search for new ideas for services for customers; shops use computer-controlled ordering and stock management; farms use new seeds, fertilizers, and irrigation technologies, satellites can assist the optimization of their planting and harvesting, and new uses are being made of their products, such as biofuels and health-promoting functional foods. Innovation is found in construction, in new materials and building techniques; in packaging that keeps food fresher; and in clothing firms introducing new designs more quickly and cheaply. Innovation is sought by public services, in health, transportation, and education. While one might not wish for too much innovation in some areas, such as amongst the firms investing our pension funds or designing the aeroplanes in which we fly, generally the business or organization that doesn't benefit from using new ideas is rare indeed.

Challenges

The challenges of innovation are immense. Many people are uncomfortable with the change brought about by innovation. Especially when it is broad-ranging, innovation can have negative effects on employees, inducing uncertainty, fear, and frustration. Organizations have social contracts by which their members develop loyalty, commitment, and trust. Innovation can disrupt this contract by redistributing resources, altering the relationships between groups, and emphasizing the ascendancy of one part of the organization to the disadvantage of others. It can disturb the technical and professional skills people acquire over many years, and with which they strongly identify themselves. Its organizational context means it is inseparable from the exertion of power and resistance to it.

Most attempts at innovation fail. History is littered with unsuccessful attempts to apply the – often very good – new ideas of individuals and organizations. The ill-fated development of a cost-effective battery electric car with significant environmental benefits in the USA in the 1990s is illustrative of the way innovation can provide a serious threat to established interests. A coalition of political and business interests combined to prevent this new idea reaching the market. Although the product was popular with consumers, it had to compete with the interests of the established energy infrastructure, oil companies, and petrol distribution networks, and massive existing automotive industry investments in petrol engine car manufacture and maintenance.

Organizations simultaneously need to do things that allow them to operate in the short term, exploiting their existing know-how and skills, and explore new things that will develop capacities to support their continued long-term existence in a changing world. These demand different and sometimes conflicting behaviours and practices. Indeed, organizations are occasionally confronted with the paradox of needing to apply new ideas threatening to the practices that have created their current successes. If generals are said to fight the last rather than the current war, managers rely on ways of doing things that contributed to their organizations', and their own, past progress, rather than ways that will deal more effectively with the future. Since Edison established the first organization dedicated to producing innovations at the turn of the 19th century, many different ways of organizing the creation and use of ideas have periodically been favoured. As the business environment has changed, the large, centralized corporate Research and Development (R&D) laboratory, and the distinctly separate innovation team (sometimes called the 'skunkworks'), are no longer used as often as in the past. The search for ways of balancing routines with innovation is constant.

Organizations rarely innovate alone: they do so in association with others, including their suppliers and customers. They innovate in particular regional and national contexts. Access to innovation-supporting skills and university research, for example, often has a local dimension, as seen in the case of Silicon Valley in California and other international centres of innovation. Government policies and regulations affect innovation, as do national financial and legal systems that influence issues such as the availability of risk-taking investment capital, the creation of technical standards, and protection of intellectual property rights. Availability and cost of infrastructure for communications and transportation matter greatly. These factors add to the complexity, and hence unpredictability, of innovation, as innovators are never completely masters or mistresses of their destiny. They also point to the essentially idiosyncratic nature of innovation: each innovation occurs in its own particular set of circumstances.

In all the major elements of contemporary economies – in the services, manufacturing, and resources industries, and in the public sector – organizational progress depends upon owning or accessing and using knowledge and information. Being competitive and efficient relies on being innovative with all the resources organizations possess: their people, capital, and technology, and the ways they connect with those contributing to and using what they do.

Innovation thinking

The American economist William Baumol argues that virtually all of the economic growth that has occurred since the 18th century is ultimately attributable to innovation. The successful application of ideas has been recognized within industry as the primary source of its development since this time.

The 18th century also saw the beginning of the study and recognition of the importance of the relationships between organization, technology, and productivity with the publication of Adam Smith's *Wealth of Nations* in 1767. Smith produced his now famous analysis of the importance of the division of labour in a pin factory, which was so influential on Wedgwood's factory organization. Smith showed how specialization in specific manufacturing processes in pin production significantly increased the productivity of the workforce compared to when individuals produced each pin themselves. A man alone, even with the 'utmost industry', could produce 1 to a maximum of 20 pins a day, yet with a division of labour, 'very poor' labour 'indifferently accommodated with the necessary machinery', could produce 4,800 'when they exerted themselves'.

A century later, Karl Marx was highly aware of the significance of innovation, but more concerned with its negative consequences. In the first volume of *Capital*, he declared:

> Modern industry never views or treats the existing form of a production process as the definitive one ... By means of machinery, chemical processes and other methods, it is continually transforming not only the technical basis of production, but also the functions of the worker and the social combinations of the labour process.

The possibilities of technological change, Marx argued, were contradicted by its use under capitalism, which inevitably led to the suppression of workers. Capitalism, he contended, subordinated workers to machines, but he believed technology held the possibility of liberating them from the burden of mechanical and repetitive work and enriching social relations.

Marx's emphasis on the strong social dimensions to technological development and use is a recurrent theme in research into the history of innovation. Study of the development of automated machine tools in the USA, for example, illustrates how often technology is shaped by dominant social forces. The automated, or numerical, control of machine tools, such as lathes, could have been configured in various ways to give the machine's operator more or less discretion over how it was used. The technology was constructed in such a way that control resided in engineering planning offices, not with their operators. This was less economically efficient, but complied with the expectations of the major customer for the new technology, the US Air Force, and hence reflected existing power structures.

At a more aggregate level, all the past revolutions in technology – in steam power, electricity, automobiles, information and communications technology – have required enormous adjustment and adaptation in industry and society. The economists Christopher Freeman and Carlota Perez show how in history the diffusion of new technologies since the Industrial Revolution required massive structural adjustments in industry and society, and also in the legal and financial framework, education and training systems for new skills and professions, new management systems, and new national and international technical standards.

The importance of clever 'human capital' has long been recognized. Observing the development of German industry in the mid-19th century, the political scientist Friedrich List declared that national wealth is created by intellectual capital: the power of people with ideas. In 1890, the British economist Alfred Marshall noted that knowledge is the most powerful engine of production available to economies. An economic theorist who unusually kept his feet on the ground by regularly visiting companies, Marshall celebrated the importance of innovation and is especially remembered for his analysis of the benefits of the 'clustering' of progressive firms in 'industrial districts'.

If any economist lays claim to be the first to include innovation centrally within their theory of development, it is Joseph Schumpeter (1883–1950), who remains today as one of the most influential thinkers on the subject. A complex man with a rich history, including being one time Finance Minister in Austria, director of a failed bank, and Harvard professor, Schumpeter argued that innovation unleashed the 'gales of creative destruction'. It arrives in a great storm of revolutionary technologies, such as oil and steel, that fundamentally change and develop the economy. Innovation is creative and beneficial, bringing new industries, wealth, and employment, and at the same time is destructive of some established firms, many products

2. **Schumpeter placed innovation centrally in his theory of economic development**

and jobs, and the dreams of failed entrepreneurs. For Schumpeter, innovation is essential for competitive survival:

> the competition from the new commodity, the new technology, the new source of supply, the new type of organization ... competition which commands a decisive cost or quality advantage and which

strikes not at the margins of the profits and the outputs of the
existing firms but at their foundations and their very lives...

Schumpeter's views on the primary sources of innovation altered
during his lifetime, reflecting changes in the practices in industry.
His early 'Mark I' model, published in 1912, celebrated the
importance of individual, heroic, risk-taking entrepreneurs. His
'Mark II' model, by contrast, published 30 years later, advanced
the role of the formal, organized innovative efforts in large
companies. It was during this period that the modern research
laboratory became firmly established, initially in the chemical and
electrical industries in Germany and the USA. By 1921, there were
more than 500 industrial research laboratories in the USA.

Five models

One of the first and most influential studies of the relationship
between scientific progress and industrial innovation was
conducted immediately after the Second World War by Vannevar
Bush, the USA's first presidential science advisor. In his report
Science: The Endless Frontier, Bush advocated a national policy for
open-ended research on a massive scale. The book proved popular;
it was serialized in *Fortune* magazine, and Bush appeared on the
front page of *Time*. The view that investments in research held the
solutions to most seemingly intractable problems was a legacy of
Bush's association with the Manhattan Project to develop the atom
bomb, which, to the minds of many, successfully curtailed the war
in the Pacific. Although it took a simplistic interpretation of Bush's
report, the view that all product and process innovations are
founded in painstaking basic research became the fundamental
precept of the 'science push' model of innovation, a perspective
that remains popular with many in the scientific research
community to this day.

An alternative view, which emphasized the importance of market
demand as the primary source of innovation, emerged in the 1950s
and 1960s. This resulted from a number of factors, including

studies that showed that in sectors such as the military, technological outcomes resulted more from the demands of its users rather than any scientifically predetermined configurations. At the same time, there was a growth of large, corporate planning offices with belief in the conceit that sufficient market research could identify what was required of new science and technology to meet consumer needs. This mirrored the rise of social science at the time with its claims to predictive powers. Counter to the post-war enthusiastic embrace of science and technology, social movements – such as Ralph Nader's car safety campaign in the 1960s, developed in response to dangerous car designs – began to question the use to which they were being put and demand greater attention to consumer needs. In housing, research into the demographics of the baby-boomer generation led to 'predict and provide' strategies internationally, where innovation was sought to help satisfy growing demand. This view became known as the 'demand-pull' model of innovation.

Both these models of innovation are linear in their progression: research leads to new products and processes introduced into the market, or market demand for new products and processes leads to research to develop them. But increasing volumes of research conducted in the 1970s questioned the assumption of linearity. Pioneering studies, such as Project SAPPHO at the University of Sussex, UK, found differences between sectors: for example, the chemical industry innovated differently from the scientific instruments industry. And, furthermore, the pattern of innovation changed over time. Abernathy and Utterback at MIT developed the theory of product life cycles, with high levels of innovation in the development of products occurring initially, then reducing in scale and being replaced by high levels of innovation focusing on their application and their processes of production. Innovation was seen not to be unidirectional, but more iterative, with feedback loops.

The organizational and skills issues underlying this 'coupling' model of innovation came to the fore in the 1980s, driven primarily by the remarkable success of Japanese industry. A study of the car industry at the time showed Japanese auto manufacturers were twice as efficient as their international competitors in every measure of innovative performance, such as how long it takes to design and make a car. The explanation for this was an approach described as 'lean production', which contrasted with the mass-production techniques used in other countries. Mass production, typified by Henry Ford, is based on assembly lines producing standardized products. 'You can have any colour Model T car you want, as long as it is black', as Ford is reputed to have said. Lean production introduced greater flexibility into the assembly line, allowing a broader range of products to be made. It includes a system of relationships with suppliers of components that allow them to deliver 'just-in-time' to be assembled, thereby reducing the cost of holding inventory, and increasing the speed of response to market changes. Lean production also entailed an obsessive concern for quality control, which in many areas became the responsibility of shop-floor workers.

When analysing the differences between the way Japanese and Western firms organized themselves to innovate, the metaphors were used at the time of the former playing rugby (although netball is similarly suitable) and the latter running a relay race. In the West, innovation entailed one part of the organization, say R&D, beginning the process, running with it for a while, then handing it over to another, say Engineering, which similarly worked on it before passing it over to Manufacturing and then on to Marketing. This linear process was considered hugely wasteful to Japanese firms, with the likelihood of significant miscommunication as projects moved from one part of the organization to another. A rugby or netball player metaphor can be used, as their games involve the simultaneous combination of different kinds of players, with various skills and abilities, some big and tall, but generally slow, and some smaller, skilful, and fast, all working to the same objective. All parts of the organization were combined in innovation activities.

Collaboration between, as well as within, innovative Japanese companies was a feature of their 1980s' success story. As well as the extensive collaboration between customers and suppliers in the same industrial groups – Keiretsu – the Japanese government also encouraged collaboration between competing firms. The Fifth Generation Computer programme, for example, attempted to encourage the usually highly competitive computer manufacturers to cooperate around shared research agendas. This 'collaborative' model of innovation strategies and public innovation policies was also enthusiastically pursued in Europe in information technology and the USA in semiconductors.

By the 1990s, Roy Rothwell, one of the founders of innovation research, began to identify a number of changes occurring in the strategies firms were using to innovate and the technologies they were using to support it. He argued firms were developing innovation strategies that were highly integrated with its partners, including 'lead customers', demanding users, and co-developers of innovation. Also important, he contended, was the use of new digital technologies, like computer-aided design and computer-aided manufacturing, that brought different parts of the firm together when developing innovations, and helped link external parties into internal development efforts. Rothwell called this the 'strategic integration and networking' model of innovation. The trend towards greater strategic and technological integration in support of innovation is one that continues with the use of massive computing power, the internet, and new visualization and virtual-reality technologies.

These models of the innovation process have their antecedents in an industrialized economy where innovation predominantly occurs in the manufacturing industry. We are now in economies where services dominate, accounting for around 80% of Gross Domestic Product in most developed nations. Economies based on tangible, physical objects that can be measured and seen have been transformed into ones where outputs are weightless and invisible. Furthermore, as the global financial crisis that emerged in 2008 shows, we live in an era of extraordinary turbulence and uncertainty where any established formulae and prescriptions are likely to be tested by emerging and unforeseen circumstances. The models of innovation in the future will be far more organic and evolutionary where the sources of innovation are unclear, the organizations involved initially unknown, and the outcomes highly constrained by unpredictabilities. In these circumstances, it will be valuable to assess whether or not anything we know of the past might be a guide to the future. It will also be useful to understand how theory of innovation might help.

Theory

There is no single, unified theory of innovation. There are partial explanations from, for example, economics, political science, sociology, geography, organizational studies, psychology, business strategy, and from within 'innovation studies', which draws on all these disciplines. This is to be expected given innovation's multiple influences, pathways, and outcomes. The utility of various theories will depend on the particular issues being explored. Theories from psychology may be more useful when the subject is an individual innovator, business strategy when it is organizational innovation, and economics when it is national innovation performance. It is important to consider theories of innovation not only to explain contemporary issues, significant as they are, but also to enlighten its future use in helping to deal with major social, economic, and environmental concerns.

Over recent years, there has been an emergence of a number of perspectives that share common ground in their theories of innovation. These include evolutionary economics and 'dynamic capabilities' frameworks for business strategy.

The challenge for any theory of innovation is that it has to explain an empirical phenomenon that takes many guises. It also has to encompass its complexity, dynamism, and uncertainty, often compounded by the way innovation results from the contribution of many parties with occasionally divergent and not fully established agendas. In this way, innovation has emergent properties: it results from a collective process whose outcomes may not be known or expected when it begins.

Evolutionary economics – with a Schumpeterian legacy – sees capitalism as a system that produces continuous variety in the new ideas, firms, and technologies created by entrepreneurs and the innovative activities of research groups. Decisions by organizations, consumers and governments make selections from within this variety. Some selections are successfully propagated and are fully developed into new organizations, businesses and technologies that provide the basis and resources for future investments in creating variety. Much of the variety and selections that occur are disruptive or fail to be propagated, so the evolutionary development of the economy is typified by significant uncertainty and failure.

Dynamic capabilities theory includes the ways firms search for, select, configure, deploy, and learn about innovations. Its focus is on the skills, processes, and organizational structures that create, use, and protect intangible and difficult to replicate assets, such as knowledge. This approach to strategy reflects the continual dynamism of technology, markets, and organizations where the capacity to sense threats and realize opportunities – when information is constrained and circumstances unpredictable – is the key to sustainable corporate advantage.

These theoretical explanations for innovation embrace complexity and emergent circumstances. They incorporate the messy organizational realities of innovation found in economies where there is constant change and adaptation and when the strategies of firms are often experimental.

Time

There has to be a time dimension to any understanding of innovation. Whether considering outcomes, what happened, or processes of innovation, how it came about, it is necessary to know the period in which they occurred. Comparisons to what existed before the innovation determine the extent of novelty.

If an innovation is ahead of its time, as perhaps might be argued to have happened in the case of the battery electric car, no matter how much effort is expended, it will not gain the momentum needed for its wide diffusion and sustained growth. If an innovation takes too long to be developed, it may fail because a superior or cheaper idea emerges. Sometimes markets and technologies shift quickly, and rapidly move on from what at one point seemed like a good idea. Innovative organizations therefore have to think about timescales of new ideas. They can do this by considering their position based on past innovation diffusion patterns, and use tools and techniques to speed up innovation through formal project management techniques that progressively decide on the levels of resources needed. Returns on innovation investment are planned over periods of years, and decisions are made to invest if they pay back suitably over an acceptable time period. Risk is managed by attempts to reduce how long it takes to develop and introduce innovation. Speed is usually, but not always, seen to be a benefit. Compressing time reduces the chance of being caught up by competitors or wastefully squandering resources. Moving too quickly, however, leads to mistakes and failures to learn from them.

Short-term horizons are appropriate for incremental innovation, but long-term perspectives are needed to provide a wider view on where, why, and how radical innovation has occurred or failed. Understanding the relationships between scientific discovery, innovation, and societal changes requires deep historical interpretation.

Innovative organizations – as we shall see in the case of Edison in Chapter 5 – can improve their future chances of success by creating options that allow different potential paths to be followed, delaying decisions that do not need to be made until a later date when their consequences may be clearer. By organizing and equipping themselves for unforeseen eventualities, innovators can change course, or re-calibrate timetables. As Louis Pasteur observed about scientific discovery through experimentation, 'chance favours the prepared mind'.

Rates of innovation and diffusion vary considerably between different business sectors. In pharmaceuticals, for example, it usually takes between 12 and 15 years to bring a new drug to market, but new digital services can grow large within months. Organizations can make strategic choices on whether they should attempt to lead innovation in their sector, or follow others. Sometimes leaders have the best opportunity for reaping the greatest rewards from their ideas. The chemical company DuPont, for example, has consistently led other firms in bringing new products to market for over a century. But 'first-mover advantage' may be difficult to capture and sustain. It often brings greater risks, as the market may not be fully formulated, and higher costs may be accrued to stimulate demand.

Other organizations choose to learn from leaders, emulating innovations that appear to work well, and avoiding any pitfalls they have observed. Fast-followers can receive huge rewards, as Microsoft has done for consistently reacting quickly to the innovations of others who have borne initial risks. Many organizations do not have the skills or resources to be first-movers or fast-followers. They may, nevertheless, benefit from innovation that improves upon, adapts, or extends products, processes, or services. Whatever its position as an innovator, and whichever strategy an organization may wish to pursue, its ability to appreciate the time dimension is likely to have a significant bearing on its performance.

Chapter 3
London's wobbly bridge: learning from failure

Schumpeter's analysis of innovation being a process of creative destruction implies the outcomes of innovation can simultaneously be positive and negative. It both creates and destroys wealth and jobs. Innovation profoundly affects us all by creating new industries, firms, and products, as seen in the new industry established by Wedgwood. It is seen in services such as discount airlines, and infrastructure such as airports. It improves productivity, and quality of life in the form, for example, of new pharmaceuticals, means of transportation, communication, entertainment, and greater variety in and accessibility of food. It has helped lift millions out of poverty, especially over recent decades in Asia. Jobs can be more creative, interesting, and challenging as a result of innovation. But the successful application of ideas can also have profoundly adverse consequences. Nations and regions get left behind when they are not as innovative as their competitors, and increasing disparities in wealth result. Jobs can be deskilled, job satisfaction decreased, and unemployment increased, because of innovation. Innovation has given us the environmental consequences of the internal combustion engine and chlorofluorocarbons, and the toxic results from the complex financial instruments behind 2008's global financial crisis.

Predicting the adverse consequences of innovation can be as challenging as foreseeing its positive effects: these are unpredictable and can be mixed. On the positive side, the internal combustion engine democratized travel, chlorofluorocarbons in refrigerators improved nutrition, financial innovations gave us the security of better life insurance and pensions. But the occasionally ambiguous nature of the results of innovation is seen in the case of failure. Most attempts at innovation fail, and there is a highly skewed distribution in its returns, but failure is itself an important outcome, and it is to this that we turn.

Failure

Innovation is risky, as, for example, innovators have to consider:

- Demand risk – how big will the market be for a new product or service? Will new competitors emerge?
- Business risk – is appropriate finance available to meet the costs of innovation? What effect will an innovation have on organizational reputation and brands?
- Technology risk – will a technology work, is it safe, and how does it complement other technologies? Will better competing technologies emerge?
- Organization risk – are the right management and organizational structures being used? Are the necessary skills and teams available?
- Network risk – are the right collaborative partners and supply chains in place? Are there important gaps?
- Contextual risks – how volatile are government policies, regulations, and taxation, and finance markets?

In theory, risk can be measured and managed by making assumptions on probabilities, although there are always dangers in assuming the past can predict the future. Uncertainty, on the other hand, has a truly unknown outcome and cannot be measured, and its management depends on decisions based on deep experience and intuition. It is because of risks and uncertainties that there is so much failure in innovation, but at the same time, they provide an incentive. If there were no risk and uncertainty, and therefore everyone could innovate easily, then innovation would provide little advantage over competitors.

Failures also provide valuable opportunities for future improvements, as seen in the highly embarrassing case of London's Millennium Bridge. Linking the Tate Gallery and St Paul's Cathedral, this was the first footbridge to be built across the River Thames for more than 100 years. It is an extraordinary engineering, architectural, and sculptural achievement – a design of such beauty it has been described as a 'blade of light' across the river. The bridge opened on the 10 June 2000, when between 80,000 and 100,000 people walked across it. When large groups of people were crossing, however, it became noticeably and increasingly unsteady, and it quickly gained notoriety as the 'wobbly bridge'. The bridge was closed after two days, causing immense discomfiture to all concerned.

After an intensive international effort, the cause was found and rectified. The problem, apparently, was the way men tend to walk with splayed feet, like ducks. When many of them walk in unison, unusual 'lateral excitation' occurs. Had it been a women-only bridge, there would not have been a problem. As a result of this debacle, new knowledge about bridge design was developed, and future projects will allow large numbers of men to waddle happily together over rivers.

The Millennium Bridge is an example of the way in which much progress in science, engineering, and innovation is built upon failure. As the chemist Humphry Davy said: 'The most important of my discoveries has been suggested to me by my failures.' And as Henry Ford put it: 'Failure is only the opportunity to begin again more intelligently.' Empirical evidence shows how returns to new ideas are highly skewed – there is what physicists and

3. The Millennium Bridge: a great success after a wobbly start

economists call a 'power law distribution'. Only a few academic papers, patents, products, and company start-ups are successes. In most cases, the majority of returns come from 10% of innovative investments. In some areas it is even more skewed. At any one time there may be up to 8,000 potential new pharmaceuticals being researched around the world, but maybe only one or two will prove successful.

There is a strongly temporal element to failure: things deemed failures can become successful, such as the Millennium Bridge, and successes can over time turn into failures. Following its introduction in 1949, the de Havilland Comet aircraft was instrumental in creating the international commercial airline industry. The Comet was considered a highly successful product innovation until the period in the mid-1950s when the aircraft started falling out of the sky with alarming regularity. Aircraft engineers at the time knew little about metal fatigue, which was the cause of the crashes, but aircraft design improved as a result of the lessons learned from these failures.

34

Products may succeed technologically but fail in the market. The Sony Betamax was technically a better video recorder than its competitor, Matsushita's VHS system, but it lost the competitive battle to become the dominant design in the market. Concorde was a technological marvel in its time, but it sold only to the British and French governments, its joint manufacturer.

It is not always possible to judge what is going to be valuable in the future. Apple's Newton – an early personal digital assistant – is a notorious product failure. It cost more than a computer, and one memorable technical review said it was so big and heavy it could only be carried around by kangaroos. Its failure cost Apple's CEO his job. Yet ten years later and its operating system was found in the iPod, and several Newton-like features were incorporated in the iPhone.

Failure has a personal cost, and innovators have to develop strategies to deal with failure that entail personal recognition of its value for learning, reflection, and self-awareness. Similarly, organizations need to appreciate the value, and learn the lessons, of failure.

Learning

Innovation is manifested in new products, services, and processes. Less material, but no less real, are the options for the future it provides and the organizational and personal learning it encourages.

Organizations learn to do better the things they already do, learn to do new things, and learn about the need to learn. Organizations inevitably learn by doing familiar things; the more you do something, generally the better at it you become. But radical and disruptive innovations – those that involve significant breakthroughs and fracture past ways of doing things – pose great difficulties for organizations and the ways they learn. Established routines and ways of doing things actually restrain learning about these forms of innovation. Focus on the status quo produces returns that are positive, proximate, and predictable; focus on the novel produces returns that are uncertain, distant, and often negative. This produces a tendency to substitute exploitation of known alternatives for the exploration of unknown ones. Radical innovation involves technologies that are destabilizing of existing capabilities, and disruptive innovations entail disengagement with existing customers and secure income streams. There are compelling reasons why organizations try to avoid them.

This is where leadership comes in, providing the encouragement and resources to do things organizations find hard, but which are necessary for their continuing viability. Positive affirmation of the outcomes of innovation, through reviews and post-project assessments, and their wide communication throughout the organization builds support for new forms of learning. When positive results from innovation become remembered as organizational stories and corporate myths, they assist efforts to break with routine and institutionalized practices, and stimulate learning in all its forms.

Employment and work

There is a continuing debate on the impact of innovation on employment and its effect on the quantity and quality of jobs. Innovation has contributed to the massive historical shift in aggregate employment from agriculture to industry to the service sectors, but its impact on industries and organizations depends on their particular circumstances and choices.

The debate itself has a long history. Adam Smith would argue that increases in market size lead to greater opportunities for the division of labour, the replacement of people with machines, and potential deskilling. For Marx, automation inevitably led to labour replacement, wage reductions, and greater oppression of the workers. Schumpeter would argue that, as innovation both creates and destroys jobs, there will be mismatches between jobs and skills in declining industries and regions and emerging innovative new sectors, and painful adjustments will be needed with periods of skill shortages and unemployment.

One view has product and service innovation producing positive effects on jobs and skills, and process and operations innovation producing negative effects. As we shall see in Chapter 5, Edison created highly skilled jobs in his 'invention factory' and large numbers of unskilled jobs in his production factory. The skilled jobs were allied to product innovation, where thinking was at a premium, the unskilled jobs were linked to process innovation, where machinery reduced the need for thinking. But there is value in having skilled workers on production lines, and organizations often make choices about how they use innovations. The way machinery is designed and tasks configured affects the use of skills. Because of these choices, and as a result of the adjustments necessary as industries evolve in response to innovation, there are major incentives for individuals, employers, and governments to invest in education and training.

Organizations need to understand how innovation can be personally rewarding and stressful, stimulating, and frightening. It can provide incentives and motivation and fear of change and loss of status. It can be divisive, with one part of the organization undertaking well-remunerated and satisfying work and others poorly paid and dissatisfied. It may be exclusive, denying people without a particular type of education, or in some cases women, access to jobs.

Economic returns

Productivity, the index of outputs to inputs, increases when resources are used more efficiently. Improvements in the use of labour and capital increase productivity. It also increases when innovation, and improvements in technology and organization, contribute to what is known as multi-factor productivity (MFP). Ultimately, economic wealth depends upon improved productivity, and this is frequently driven by innovation. MFP growth in the USA in the 1990s, for example, was linked to the information and communications industry and the use of its products in other sectors of the economy. Much recent growth in MFP has occurred in the service industries, such as retailing and wholesaling, and this can partly be attributed to the use of digital technologies.

Profitability is driven by a wide number of factors such as how much better and efficient organizations are compared to competitors at designing, making, and delivering things, and the preferences of customers for particular brand names and their preparedness to pay prices that provide the required return to innovators. Innovation contributes to profits by providing distinctive advantages in the sale of products and services; in their features, prices, delivery times, upgrade opportunities, or maintenance. Intellectual property can be sold and licensed, and new start-up businesses can be created, to produce profit from innovation. Large-scale innovative activity, in investments in R&D or plant and equipment, can deter competition and hence improve opportunities for profit.

For organizations to benefit financially from investments in innovation, they have to appropriate returns. In some circumstances, innovation can be protected by using intellectual property law for patents, copyrights, and trademarks. In others, protection derives from difficult to replicate skills and behaviours, such as the capacity to quickly stay ahead of competitors, being able to maintain secrecy, or retaining important staff. In all cases, the contributions of innovations to profits are frequently skewed, with most returns from a few innovations.

Technical standards that allow inter-operability between components and systems confer economic advantage. Those organizations that own standards, or whose offerings comply with them, have advantages over those that do not. Battles over technical standards can become particularly heated, as we shall see in the case of Edison in Chapter 5.

Chapter 4

Stephanie Kwolek's new polymer: from labs to riches

Many people and organizations contribute to innovation. Large-scale surveys of innovative firms, such as the European Union's Community Innovation Survey, for example, show a wide range of contributors. These surveys also rank the importance of various sources, showing the most important of them lies within the organization. Innovation derives primarily from the energy, imagination, and local knowledge of employees identifying and solving problems. It is stimulated by innovative individuals and workplaces, and by formal organization structures and practices, such as R&D departments and management tools for developing new products.

Second in importance as sources of innovation, according to these surveys, are customers and clients, followed by suppliers of goods and services. Fairs and exhibitions, professional conferences and meetings, and academic and trade journals are reported to be important by a minority of firms. The least important sources, these surveys show, are universities and government research laboratories.

These rankings hide a much more complicated picture. Reliance on internally derived innovation, for example, makes organizations introspective and perhaps unprepared to deal with changes occurring externally in markets and technologies. Depending on customers for innovative ideas will likely produce conservative 'don't rock the boat' approaches. Universities are critically important contributors to invention for science-based sectors, and for innovative products and services at early stages of gestation, and they also educate and train employees with the skills to innovate.

As Josiah Wedgwood showed us, innovation usually involves the combination of ideas derived from many different starting places. The great scientist Linus Pauling said the best way to have a good idea is to have lots of them, and the same sentiment can apply to the pursuit of innovation from multiple contributors. Schumpeter's contention that innovation requires 'new combinations' amongst markets, technologies, and knowledge often entails integrating ideas from many different parts of the organization and with various external parties. The stimulus to innovate may not result from particular sources, with hierarchical contributions, but from multiple sources of ideas that intersect and blur in circumstances of intrinsic necessity and the impulsive quest for survival in volatile times.

Innovation is also affected by wider social, cultural, political, and economic factors. These include the contributions made by cities and regions, government policies, and the 'systems of innovation' organizations belong and contribute to.

Continual pursuit: the case of IBM

The continual, broad-ranging, and challenging pursuit of innovation is seen throughout the history of the IBM Corporation. IBM is widely recognized as one of the most innovative companies in the world, playing a central role in the discovery and development of, amongst other things, supercomputers, semiconductors, and superconductivity. It has invested enormous resources in innovation. It spends billions of dollars on R&D each year, produces more patents than any other company, regularly creates iconic products and services, and its staff have won five Nobel Prizes. It has huge advantages in innovation compared to almost any other company in the world, yet its pursuit of innovation holds lessons for other organizations.

IBM was incorporated in 1924, but its history can be traced back to Herman Hollerith's foundation of the Tabulating Machine Company in 1896. Hollerith (1860–1929) developed a machine using electricity and card processors to mechanize the data-processing of US Census data. He called the machine 'hardware' and the cards 'software'. Hollerith worked for a time at the US Census Bureau and was acutely aware of its need for improved efficiency in processing data. The 1880 Census had taken seven years to compile, and there were fears that the 1890 version was going to take longer. Hollerith's tabulating machine met the Census Bureau's requirement for fast and efficient data collection and management. By using it, the 1890 data were analysed in six months, saving millions of dollars, and it was subsequently used in censuses in Canada and Europe. By 1912, Hollerith had sold his business and, although he remained as chief consulting engineer, he had less and less association with the company. For many years, he had refused to respond to requests and ideas from the Census Bureau for improvements to his machines. When Hollerith's main patents expired in mid-1906, the Bureau developed a tabulator of its own, which it used in the 1910 Census. It took the arrival of Thomas Watson in 1914 to improve the technical performance of tabulating machines and improve the company's relationships with its customers.

As Chairman of IBM, Thomas Watson (1874–1956) was instrumental in developing the company's use of electronics. He underwrote Harvard scientist Howard Aiken's 1930s research into developing a digital calculating machine. In 1945, in collaboration with Columbia University, he opened the first Watson Scientific Computing Laboratory, in New York. The IBM Thomas Watson laboratory remains today as one of the largest industrial research laboratories in the world. During the Second World War, the company developed very close relationships with the US government, especially in military ordnance and planning wartime logistics. It limited its profit margins for military work to contribute to the war effort.

During his 42 years at IBM, Watson built the company into a major international corporation. His son, Thomas Watson Jr, succeeded him as Chairman. From the late 1950s to the 1980s, following massive investments in R&D, IBM became the world leader in mainframe computers, especially with its System 360, launched in 1964. The System 360 remains in real terms one of the largest private investments ever made in R&D. The company, which was valued at $1 billion at the time, committed $5 billion to its development. Tom Watson Jr had 'bet the

4. The IBM System/360 computer. IBM 'bet the company' on its development

company' on its development. By 1985, IBM had 70% of the world mainframe market. It had unmatched expertise in hardware and software, and its business skills made it one of the world's most admired companies.

By the mid-1970s, the company began the pursuit for smaller computers. The IBM Personal Computer (PC), launched in 1981, was along with the System 360 one of the most iconic products of the last century; it essentially created the mass market for PCs. It emerged from an IBM development group in Boca Raton in Florida that had failed in three previous attempts to create a PC. The successful development of the PC required the rejection of IBM's past strategy of self-reliance and developing everything itself internally. It decided to buy major components such as integrated circuits and operating software from small suppliers. Initially, the product was a huge success, capturing 40% of the market.

By the late 1980s and early 1990s, however, IBM was in serious trouble and nearly went bankrupt. The IBM PC had helped sow the seeds of its own demise. IBM did not control the intellectual property rights to its components, and the small suppliers – Intel and Microsoft – grew quickly to be larger and more powerful than IBM and supplied its competitors with their technology. Furthermore, the overall culture of IBM remained focused on historically profitable mainframes at the same time as price competition from Japanese manufacturers led profit margins to collapse. *The New York Times* of 16 December 1992 was led to the opinion in an editorial that 'The IBM era is over ... what was once one of the world's most vaunted high-tech companies has been reduced to the role of a follower, frequently responding slowly and ineffectively to the major technological forces shaping the industry.' The story of Herman Hollerith's rise and fall resonated again.

One response to IBM's 'near-death' experience was to appoint a new CEO, Lou Gerstner, the first appointed from outside IBM.

The company went through a massive restructuring and fundamental shift in business strategy. It changed from being a supplier of technology to being a provider of solutions to customer problems. Its objective was to provide the best possible service for customers, even if this meant using competitors' technologies. At the same time, despite its financial difficulties, the decision was made that as the company's strength in the past derived from its 'science and engineering mindset', its research investment should continue in the future. The search was on to find increasing innovations from within the company's technological community and R&D centres. These internal sources essentially reinvented the mainframe using microprocessors and parallel architectures. IBM also became much more open to ideas from outside, attempting to break away from its past introspection and 'not-invented-here' syndrome. It began to use open technical standards and software rather than those it owned proprietarily and started to collaborate more in its technological developments, embarking annually on scores of collaborations with other organizations. Its new 'market-facing' innovations included supercomputing, e-business, social networks, and Web 2.0 technologies.

The company presently makes extensive use of its intranet and social networking technologies to access and share ideas amongst its staff. With around 400,000 employees, half of whom are scientists and engineers, and 75 research centres globally, the company has massive technological skills on which to draw. The ways it uses these skills to support its current and emerging innovation process will be discussed in Chapter 6.

IBM illustrates the extensive pursuit of innovation throughout its history, involving inventor-entrepreneurs, customers, suppliers, universities, R&D departments, relationships with government, collaborative partners, and the broad community of its own staff and contacts. We now turn to these various contributors.

Entrepreneurs and venture capitalists

In contrast with the large-scale activities of companies such as IBM, innovation also results from individual entrepreneurs using it to build new businesses. The term 'entrepreneur' began to be used in the early 18th century and is applied to individuals who discover, recognize, or create opportunities and then manage resources and bear risks to take advantage of them. Wedgwood demonstrates the substantial contribution entrepreneurs can make to innovation and economic development.

From Matthew Boulton in the 18th century, to Thomas Edison in the 19th, to Bill Gates in the 20th, and Sergey Brin and Larry Page in the 21st, entrepreneurs are commonly associated with the creation of technology-based companies. These companies grow rapidly on the basis of novel technologies that create new industries and transform old ones. Some entrepreneurs transform entire economies and societies. Boulton and his partner, James Watt, developed the steam engine and the world's first mechanized factory and helped usher in the Industrial Revolution. Edison, amongst many other contributions, developed electric power-generation technology and created the General Electric Company. Software from Gates' Microsoft popularized the personal computer; Brin and Page's Google transformed the use of the internet; and both companies have changed the nature of work and leisure.

These are highly exceptional examples. Around half a million new companies are formed annually in the USA alone, and very few, if any, will be as successful as Microsoft and Google. Yet the creation of new firms, and the challenges they present to incumbent firms, is an essential element and a major contribution of capitalism. In Schumpeter's Mark I model, creative destruction is driven by the entrepreneurial task of 'breaking up old, and creating new tradition'.

Schumpeter on entrepreneurs: some selected quotations

Some of the many motivations and characteristics of the entrepreneur identified by Schumpeter have resonance today.

Motivation:

'the dream and will to found a private kingdom, usually, though not necessarily, also a dynasty';

'the will to conquer: the impulse to fight, to prove oneself superior to others, to succeed for the sake, not of the fruits of success, but of success itself';

'the joy of creating, of getting things done, or simply exercising one's energy and ingenuity';

'private property ... pecuniary gain ... (and) other social arrangements not involving private gain'.

Character:

The entrepreneur...
'seeks out difficulties, changes in order to change, delights in ventures';

needs: 'extraordinary physical and nervous energy';

possesses: 'that special kind of "vision" ... concentration on business to the exclusion of other interests, cool and hard-headed shrewdness – traits by no means irreconcilable with passion';

knows how to: 'woo support' among colleagues, 'handle men (sic) with consummate skills,' and 'give others ample credit for the organization's achievements'.

Schumpeter's Mark II model recognized that entrepreneurship occurs in large firms as well as newly created firms, reflecting the changing industrial realities as formally organized, large-scale R&D activities grew in scale from the 1920s. Entrepreneurship is therefore the organizational process by which opportunities are sought, developed, and exploited in many different kinds of company.

In some circumstances, entrepreneurial start-up firms receive investments from venture capitalists that are prepared to assume higher risks than high street and investment banks. Many of the US success stories of entrepreneurial information technology and biotechnology companies, such as Google and Genentech, received venture capital. Different international models of venture capital exist, but the US is often considered exemplary. US venture capital may include funds from private investors or corporations, and their managers may possess deep experience or knowledge of particular technological sectors and become engaged in the governance of start-up companies. The objective of venture capitalists is usually to acquire shares in companies in their early years that then reap extraordinary returns after they exit when the firms have reached sufficient maturity to attract a purchaser or to be floated on a stock market. Amongst their portfolio of investments, venture capitalists recognize that the majority of returns will come from a limited number of cases. Generally, venture capitalists tend to invest in better-established, rather than new and speculative ventures, when technological and market opportunities have been clearly identified.

Research and Development (R&D)

R&D is a significant, but not always essential, source of innovation. Investments in R&D help organizations search for and find new ideas and improve their capacities to absorb knowledge from external sources.

R&D ranges from basic research driven by curiosity and little concern for its application, to highly practical problem-solving (see the definitions in the Frascati Manual below). Its expenditure reflects highly varied national, sectoral, and corporate commitments to its use in pursuit of innovation. Internationally, around $800 billion is spent annually on R&D. At an aggregate level, it is concentrated in a few main industries, including information and communication technology and pharmaceuticals. The USA is the major spender on absolute amounts of R&D. When relative expenditures on R&D are assessed – usually measured as a portion of a nation's Gross Domestic Product – smaller European nations, such as Finland, Sweden, and Switzerland head the list, committing over 3% of GDP annually. A marked trend in recent years has been the rapid growth in R&D expenditures in Asian nations, such as Korea, Taiwan, and China. Over 95% of R&D globally is spent in the USA, Europe, and Asia (primarily northeast Asia), so many nations, especially in the Southern hemisphere, cannot compete in this important source of wealth creation and growth.

There are wide differences among nations in the breakdown in R&D expenditures between that spent in business and government. In some countries, such as Korea and Japan, business expenditure predominates. In others, such as Poland and Portugal, government is the major source of R&D spending.

The Frascati Manual

In 1963, the Organization for Economic Cooperation and Development (OECD) decided it would be useful for policy-making to have consistent international data on R&D statistics. Following a meeting in Frascati, Italy, the Proposed Standard Practice for Surveys of Research and Development was produced.

This became known as the Frascati Manual. The sixth edition of the Manual was produced in 2004.

R&D is defined as comprising creative work undertaken on a systematic basis in order to increase the stock of knowledge, including knowledge of man, culture, and society, and the use of this stock of knowledge to devise new applications.

R&D is described as covering three activities:

- Basic research is experimental or theoretical work undertaken primarily to acquire new knowledge of the underlying foundation of phenomena and observable facts, without any particular application or use in view.

- Applied research is also original investigation undertaken in order to acquire new knowledge. It is, however, directed primarily towards a specific practical aim or objective.

- Experimental development is systematic work, drawing on existing knowledge gained from research and/or practical experience, which is directed to producing new materials, products, or devices, to installing new processes, systems, and services, or to improving substantially those already produced or installed.

The Frascati Manual has been useful in building consistent data sets on R&D expenditures internationally. It has continually evolved and improved. Nonetheless, significant problems remain in measuring collaborative R&D and activities undertaken in services.

The OECD has also developed the Oslo Manual to guide national innovation surveys, the Canberra Manual for measuring human resources in science and technology, and a Patent Manual on the use of patent statistics.

Stephanie Kwolek's new polymer

Stephanie Kwolek (b. 1923) has saved thousands of police officers and military personnel from death or disability. As a result of a traditional R&D process, she invented Kevlar, a fibre used in body armour. The product, which is one of the strongest fibres ever made, has over 200 applications, including in brake pads, spacecraft, sporting goods, fibreoptic cables, fireproof mattresses, storm protectors, and wind turbines. It has produced several hundreds of million dollars annually for the chemical company DuPont. It is best known, however, for its use in bullet-proof vests. In 1987, the International Association of Chiefs of Police and DuPont began a Kevlar Survivors' Club for those saved by the product from death or serious injury. Its 3,000th member was inducted in 2006. Kevlar's protective properties have also been used extensively in the military.

Kwolek was born in New Kensington, Pennsylvania. Her steelworker father died when she was young, but she retained

5. **Stephanie Kwolek, inventor of Kevlar**

his curiosity: he was a keen amateur naturalist. She remembers spending hours designing and making clothes for her dolls and being very interested in fashion. She studied at a college that became part of Carnegie-Mellon University, and, unable to afford to study medicine, she majored in chemistry.

She decided she wanted to work for DuPont. DuPont was and remains one of the world's leading and most innovative companies. In the 1920s, it was one of the first companies to invest in basic research with the 'object of establishing or discovering new scientific facts'. It developed neoprene synthetic rubber in 1933 and nylon in 1938. With the shortages of male chemists resulting from the Second World War, women were being attracted into the chemical industry. During her interview, Kwolek forcefully demanded to know when she would hear about the job as she had another offer. The offer was made that evening.

Kwolek began work for DuPont in 1946. She worked at the DuPont Research Laboratory in Delaware for 36 years, having previously worked for the same group for 4 years in Buffalo, New York. Her job was to develop new polymers and ways of making them. Shortly after her arrival, she was given the job of looking for a breakthrough fibre to be used to make tyres lighter and stiffer. There was a concern at the time to improve vehicle performance to address a petrol shortage. Others had been offered the task, but weren't interested. Kwolek felt that, while her competence was recognized by male colleagues, she was often overlooked.

She liked the working atmosphere, however, and the challenges that arose, and as one of the few women scientists at the time, she worked exceptionally hard to retain her job after the men returned from the war. She was given a high degree of independence and freedom to do what she wanted.

(She complains about modern research being too rushed
and short term with not enough time for thinking.)

Kwolek's specialization lay in low-temperature processes for the
preparation of condensation polymers. In 1964, she discovered
that the molecules of extended-chain aromatic polyamides
formed under certain conditions a liquid-crystalline solution that
could be spun into a strong fibre. She took her polymer, which
was unpromisingly cloudy and thin, to a machine to be spun.
She says the polymer had such strange features that anyone not
thinking or being unaware would have thrown it out. The
technician in charge of the spinner was deeply sceptical, thinking
his machine would become bunged up by this contaminated
substance, but was eventually persuaded to try. It spun
successfully a product that was so strong that Kwolek had to
repeat tests many times before she was convinced about her
discovery. She did not tell anyone about her polymer until
she was sure of its properties. Kevlar is heat-resistant, five times
stronger than steel, and about half as light as fibreglass.

DuPont was immediately convinced of the value of Kwolek's
new crystalline polymers, and the Pioneering Laboratory was
given the job of finding commercial applications. She provided
a small amount of fibre to a colleague experimenting with
bullet-proof armour. Kevlar was introduced for this purpose
in 1971. One of the reasons that it is used in such a broad range
of applications is its versatility: it can be converted into yarn or
thread, continuous filament yarn, fibrillated pulp, and sheeting.
The new chemistry that Kwolek developed helped DuPont
develop a range of other fibres, such as spandex Lycra and
the heat-resistant Nomex.

Kwolek attributes her success to the way she can see things
that others cannot. And she says:

To invent, I draw upon my knowledge, intuition, creativity, experience, common sense, perseverance, flexibility, and hard work. I try to visualize that desired product, its properties, and means of achieving it ... Some inventions result from unexpected events and the ability to recognize these and use them to advantage.

Kwolek has 17 patents, including 5 for the Kevlar prototype. She has won numerous prestigious awards and has spoken about the great need for recognition of scientists and other people who do things that benefit mankind. She admits to being very pleased when a police officer asked for her autograph on the jacket that saved his life.

The case of Kwolek and Kevlar epitomizes the corporate R&D department's contribution to innovation. It also points to some of its shortcomings. The polymer was 18 years in development, and its commercialization took 7 years. Few, if any, organizations have the capacity to take such a long-term approach nowadays.

Customers and suppliers

Innovations do not succeed unless customers or clients use them, and if the users of new products and services are involved in designing what they need, there is generally a better chance of success than if something is being designed for them. Demands and needs can never be articulated fully and communicated completely across organizational boundaries between producers of innovation and their customers and suppliers, and active engagement between them overcomes these barriers.

In some fields, such as medical instruments, the innovator is commonly the user of the innovation. Surgeons and medical practitioners are regular contributors of ideas for new tools and techniques that help them do their jobs better. The world's largest producer of implanted hearing devices, Cochlear, originated with Professor Graeme Clark, a medical researcher whose father was profoundly deaf. Clark was highly sensitized to the suffering of people whose deafness could not be helped by hearing aids, and was driven to improve their lives.

According to one estimate, up to one-quarter of all men over 30 suffer from sleep apnoea, a condition that causes potentially dangerous breathing irregularities when they are asleep. Respiratory medical devices can help manage the problem. The origins of the world's largest manufacturer of respiratory devices – ResMed – lay with Professor Colin Sullivan, a medical researcher working in a hospital sleep clinic. He overcame the problem by regularly blowing puffs of air up the nasal passage. Fortunately, for sufferers and their partners, as a result of continuous design improvements the present discreet and quiet devices are a big improvement from the original version built from a gas mask and vacuum cleaner.

Some companies go to great lengths to engage customers in designing new products. When Boeing developed the 777 aircraft, it involved its major customers, United, British Airways, Singapore Airlines, and Qantas, in trying to understand the demands of the market. It needed to know about optimum passenger loadings for airlines' favoured routes. But it also worked to understand the demands of the users of the aircraft: the pilots and aircrew, maintenance engineers, and cleaners. It aimed to be sympathetic to flight attendants brewing coffee in turbulence and maintenance engineers fixing an external component in Alaska at midnight in -40 degrees centigrade or Jeddah at midday in $50+$ degrees. As Boeing developed the 787, it built a website to gain immediate input into the design process from interested parties around the world. Around 500,000 people voted on the choice of the aircraft's name: Dreamliner.

Software companies sometimes release their products in 'beta' form, that is, in prototype, to allow users to play with the software and suggest improvements. Essentially, customers do much of the final polishing of the product. This strategy is pursued for products that are proprietary, when the company aims to profit from them. This is different from open-source software, such as the web browser Mozilla Firefox and the operating system Linux, which is built, maintained, and continually improved by networks of volunteer programmers.

Excluding customers from the process of product improvement can be very shortsighted. When Sony developed its robot dog – the Aibo – it kept its software code secret. A community of hackers grew up, developing a much more extensive range of moves for the robot, including a number of amusing dances that made it a much more attractive product for customers. Sony sued the hackers and closed down their community, but soon after recognized its mistake and realized that the company could learn from the software developed externally. Sony no longer produces the Aibo, but its subsequent products have benefited from the technology developed for the robot dog in areas such as visualization.

Customers can also inhibit innovation. They can be conservative and complacent and locked into ways of doing things that preclude novelty and risk. Clayton Christensen identified the 'innovator's dilemma'; the problem of listening too closely to customers. If innovators only respond to the immediate demands of customers, they often miss big changes occurring in technologies and markets that may eventually put them out of business. Here there is advantage in working with 'lead customers', governments, firms, or individuals that are prepared to take risks to promote innovation in the belief that greater benefits will accrue than pursuing the safer short-term option of not innovating. In the 1980s, Roy Rothwell described the relationship between Boeing and Rolls Royce as 'tough customers; good designs', implying that Boeing's very demanding requirements of its aero engine suppliers made Rolls Royce design and produce better products.

Innovative suppliers are also major stimulants to new ideas. In the automotive industry, a high percentage of the value of a car is bought from suppliers of components, accounting in Toyota's case for up to 70% of the car's total cost. Toyota enjoys very close relationships with Nippondenso, a very large components supplier of innovative products such as lighting and braking systems. The automotive supplier Robert Bosch plays a similar role in the European car industry. Large automotive companies use numerous methods, including websites and technical conferences and fairs, to encourage their suppliers to provide innovative solutions to the problems they face. Innovative automobiles are based on suppliers of innovative components to car companies. The task of the car manufacturer – or the organization responsible for integrating any system of different elements – is to encourage innovation in suppliers of modules or components while ensuring the compatibility of components with overall design architectures or systems.

The encouragement of innovative suppliers is also a key objective of many governments. In the USA, the Small Business Innovation Research scheme uses the government's enormous procurement budget to support small companies by purchasing innovative products and services. This particular government scheme invests more in innovation in start-up companies than the US venture capital industry, and does so in more early stages of their development.

Collaborators

Innovation rarely results from the activities of single organizations and more commonly occurs when two or more organizations collaborate. For many organizations, the benefits of using collaboration to contribute to innovation outweigh the costs of sharing the returns to that innovation. Collaborations take the form of joint ventures and various types of partnerships, alliances, and contracts that involve joint commitments to mutually agreed aims. They can be with customers and suppliers, organizations in other industries, and even with competitors. They are a feature of the world's industrialized economies, with some collaborations operating for many decades.

Organizations collaborate to reduce the costs of developing innovation, to access different knowledge sets and skills to the ones they possess, and use it as an opportunity to learn from partners about new technologies, organizational practices, and strategies. In uncertain and evolving circumstances, innovating collaboratively provides a greater chance of success than going it alone. Information, communications, and other technologies have made collaboration cheaper and easier. Governments around the world have actively promoted collaboration as a source of innovation. And organizations have become less self-reliant and more open to collaboration in their strategies for innovation.

Different types of collaboration work better in different situations. When the objectives of collaboration are clear, or the focus is on quickly reducing costs, it works better when organizations are similar. The opportunities for misunderstandings and miscommunication are fewer. When objectives are emergent, and the objective is exploration and learning, collaboration benefits from dissimilar organizations working together. More is learned from variety than uniformity. Larger numbers of partners increase the scale of effort; fewer partners improve speed.

Collaboration can be difficult to manage. Partners may have different priorities and organizational cultures. There are many opportunities for misunderstanding, as the following, perhaps apocryphal, anecdote reveals. Some years ago, a collaboration was proposed between a group of IBM and Apple staff. Prior to the first joint meeting, IBM staff discussed their approach.

Conscious of their reputation for formality – blue suits were the uniform of the day – they decided to put the Apple staff – who were usually casually dressed – at ease by wearing their weekend clothes to the meeting. They arrived in jeans and sweatshirts to find the Apple staff sitting uncomfortably in newly purchased blue suits. That this could occur amongst organizations in the same industry and country highlights the potential problems that might occur in collaborations in different sectors and nations.

Universities

Clark Kerr, renowned social scientist and president of the University of California, was highly prescient in identifying the importance of universities for economic development when he wrote in 1963 that:

> ...the university's invisible product, knowledge, may be the most powerful single element in our culture...the university is being called on to produce knowledge as never before...And it is also being called upon to transmit knowledge to an unprecedented proportion of the population.

He argued new knowledge is the most important factor in economic growth and highlighted the role of the university in developing new industries and generating regional growth, and emphasized the contribution of the entrepreneurial professor, consulting and working closely with business. In the intervening decades, universities have been increasingly encouraged by governments and business to devote their energies to actively translating their knowledge into economic activity, a policy they have often enthusiastically endorsed. This activity is now so elevated that it has emerged, according to some, to be as important a university function as research and teaching. The ways knowledge is transferred to industry, and universities contribute to innovation, are often conceived too simply, however, and the paths to market are commonly complex, multifaceted, and subtle. The idea that ideas and knowledge are something that universities produce and 'transmit' to industry has also been replaced by the notion that it is co-created and exchanged.

Teaching

By educating skilled undergraduates, graduates, and post-doctoral students, universities prepare a labour force's capacity to create and apply new ideas. The history of the successful development of new industries – such as electrical, chemical, aeronautical, and information technology – is in major part explained by the provision by universities of sufficient numbers of graduates with the requisite new skills, especially in engineering and management. It is said that the best form of knowledge exchange between university and industry is carried on two legs and by the movement of problem solvers from university to industry.

It is not only science and engineering graduates that contribute to innovation. At various times, philosophers and anthropologists have been in demand in Silicon Valley, and the creative industries provide a home for many Humanities students. Business schools increasingly provide courses in innovation management and entrepreneurship for students of all disciplines. The management literature discusses how successful firms are said to require a combination of 'I'-shaped people, with deep knowledge in a particular field, and those that are 'T'-shaped, having a breadth of knowledge with a particular specialism. The capacity to see connections 'across the T' between various disciplines is a major stimulus to innovation, but poses significant challenges for educators (see MIT and the education of engineers).

Technical colleges also play an important role in innovation, for example in training technicians to produce instrumentation which they themselves occasionally commercialize.

Massachusetts Institute of Technology and the continuing challenge of educating engineers

Engineering is problem-solving, and to encourage this ability MIT has traditionally attempted to encourage interdisciplinarity in its educational approach. In the MIT Bulletin of 1954–5, it was said that the aim of its School of Humanities and Social Studies was to develop 'the first-rate human and social values which must accompany technical competence if an individual is to make his (sic) maximum contribution as a citizen'. The curriculum aimed to reflect these values. All students in the first two years of their four-year degrees undertook core courses encompassing history, philosophy, and literature. Their focus was on problems rather than solutions, and the development of an attitude that education was something they have to use continuously and develop, rather than something they have had.

The contemporary challenges of educating 'rounded' engineers are described by Rosalind Williams, MIT's former Dean of Students. She says that engineers nowadays need to understand how things are designed and brought to market, and how organizations work and innovations become successful. Indeed, she quotes a colleague saying how MIT had long ago given up training professional engineers and were really training technological innovators. She claims they need to understand science, on the one hand, and humanities, arts, social sciences, and management, on the other, and as a result the amount of information needed to be crammed into the head of a student doubles every 18 months. To accommodate this breadth, the trend, she reports, is the separation of engineers with different psychological and sociological profiles into 'systems integrators' and 'designers'; the former more intent on managing large technological systems in established firms, the latter more interested in entrepreneurially creating new products and services.

Science and research

Science, from the Latin *scientia* – knowledge – has been a feature of human development since the first civilizations. The application of science in industrial innovation, however, only began in earnest during the Industrial Revolution and has been most particularly a feature of the last 150 years or so.

One of the traditional distinctions in research, seen in the Frascati Manual, is between that which is 'basic' and that which is 'applied'. The former is thought to be curiosity-driven, with no consideration of its application, and the particular concern of universities. The latter is believed to be directed towards an identified use, usually in industry. Yet some businesses invest substantially in basic research and universities conduct extensive applied research, especially in professional departments such as medicine and engineering.

Furthermore, as Donald Stokes argued, the classic distinction between 'pure' or basic research, driven by a desire to understand, and applied research, with the purpose to be used, fails to capture a third category which aims to do both by improving understanding and being useful. He calls this 'Pasteur's Quadrant' of use-inspired basic research (see Figure 6). Pasteur's microbiology research was always concerned with useful applications, but created a new field of scientific understanding. Stokes contrasts this with Bohr's research in physics, in which his understanding of atomic structure provided a basis for developing the theory of quantum mechanics, and Edison's research that was driven by a concern for use and profit, although he was also influenced by theory. There is a direct and obvious connection between research and innovation in Edison's and Pasteur's Quadrants: the connection in Bohr's may or may not occur, and if it materializes, it might be in unexpected or unimagined areas. Bohr, one imagines, would have had little

	No	Yes
Yes	**Pure Basic Research (Bohr)**	**Use-Inspired Basic Research (Pasteur)**
No		**Pure Applied Research (Edison)**

Quest for Fundamental Understanding?

Considerations of Use?

6. **Pasteur's Quadrant: From Donald Stokes *Pasteur's Quadrant*** (Washington, DC: 1997)

appreciation of how quantum theory is used to explain lasers and potentially provide the basis of future quantum computers.

Similarly, in their short letter to *Nature* on 25 April 1953, Watson and Crick modestly stated: 'We wish to suggest a structure for the salt of deoxyribose nucleic acid (D.N.A.) This structure has novel features which are of considerable biological interest.' They did not imagine the considerable commercial interest that was to emerge more than 20 years later, or the way in which their discovery has transformed old and created new businesses with the development of biotechnology.

In reality, basic and applied research are elements of a continuum, with many interconnections. Applied research may result from basic research findings, and basic research may be undertaken to explain how an existing technology works. One of the most useful outcomes from pure basic research is the instrumentation

63

"We wish to suggest a structure for the salt of deoxyribose nucleic acid (D.N.A). This structure has novel features which are of considerable biological interest."

"It has not escaped our notice that the specific pairing we have postulated immediately suggests a possible copying mechanism for genetic material."

7. Letter to *Nature* announcing the discovery of DNA

developed to assist experimentation. The computer, laser, and Worldwide Web were developed to this end, with little appreciation of potential industrial use as now ubiquitous innovations.

When we consider the world's most complex scientific and social questions, including global warming, sustainable energy, food security, and genetic engineering, the answers will depend on fundamental understanding developed by universities and its practical use in industry.

Engagement

Dr Jonas Salk was reputedly once asked who owned the polio vaccine he developed. His answer was, 'why the people, I would say'. Such a response is unlikely today. Since the passing of the Bayh-Dole Act in the USA in 1980, which allowed research institutions to own the results of publicly funded research, universities in developed economies have become preoccupied with making money from their research. This has usually taken the form of patent-protected intellectual property, licensed to businesses, or through start-up companies, spun out of and part-owned by the university. Evidence suggests, however, that the number of successful instances of this model of commercialization is limited. There are some impressive success stories, such as the biotechnology firm Genentech. This company was formed in 1976 to help commercialize the discovery of recombinant DNA at Stanford University, and was sold to a Swiss pharmaceutical company in 2009 for just under $50 billion. Such companies, however, are a tiny fraction of the total amount of entrepreneurial activity encouraged by universities.

The focus of the majority of attention by governments, and indeed by many universities themselves, has been on issues such as patents and licensing, contract and cooperative research, and incubation and entrepreneurship centres. These activities are important for innovation in industries that are based on science and technology, but not for all industry sectors. They are generally less relevant for services, resources, and traditional industries, such as clothing and textiles. Furthermore, they ignore the importance of the social and networking activities that are crucial to the 'conversations' between universities and business about new developments and their potential applications. Although for many businesses, particularly smaller ones, the purpose of collaborating with universities is immediate problem-solving, larger firms will engage in broader dialogue with universities to learn about the directions of future research. Businesses claim that the attraction of working with universities is that they have different cultures to their own. University staff have more time to think about and test new ideas.

As contributors to inventive ideas and knowledge creation and diffusion, universities and research institutes need continually to communicate their capabilities and assess how they should best engage with external parties. Government and business cannot be expected to invest in universities and research institutions as providers of innovation without them fully articulating their broad contributing roles.

Regions and cities

Innovation agglomerates by localizing within particular geographies, such as in the Staffordshire Potteries. It does so for economic reasons, as proximity reduces the costs of transactions and transportation, and firms in close association stimulate the creation and diffusion of innovation through improved awareness and knowledge of each other. Innovation clusters for social and cultural reasons, including the advantages derived from shared identity and higher trust in affiliated and cohesive groups. Communications are assisted by propinquity because knowledge is sticky and travels badly from its source, especially when it is complex or tacit and cannot be written down.

The best-known innovative region is Silicon Valley, near San Francisco, an area of high technology business concentration and employment, and one that has stimulated countless and often fruitless attempts at replication around the world.

A number of factors contributed to the development and growth of Silicon Valley. Government played a central role, from the gift of land to local universities to stimulate industrial development, to being a large-scale customer of high-tech goods in defence markets. Universities have contributed their research and the education and training of scientists, technologists, and entrepreneurs. Institutions such as Stanford University have proactively developed policies to encourage academic engagement with business in fields such as electronics and information technology. Numerous high-tech businesses have started up and some have grown rapidly into major corporations, such as Hewlett-Packard, Apple, and Intel, assisted by a highly skilled and mobile labour market attractive to talented employees, links to university research, and ready access to professional services such as venture capitalists and patent lawyers. These factors contribute to a local culture, or 'buzz', that is technology-focused, risk-taking, and highly competitive, and creates a virtuous circle of initiative and reward. It has created enormous wealth and extensive experience of innovation and entrepreneurship to be reinvested back into new initiatives.

It is often not regions but cities that provide the locus of innovation. Throughout history, cities have at various stages been associated with creativity and innovation, from Athens in the 5th century BC to Florence in the 14th century, to *fin de siècle* 19th-century Paris.

Cities are major contributors to the supply and demand of innovation. Most patents emanate and R&D is conducted in cities, and their higher disposable incomes ensure greater consumption of innovation. Some cities are renowned as centres of learning, such as Oxford or Heidelberg, others for their engineering ingenuity, such as Stuttgart or Birmingham, financial and services innovation, such as London and New York, and creativity and design, such as Copenhagen and Milan. Some cities are known for their particular technology expertise, such as Bangalore and Hyderabad in India, or their support for technological entrepreneurship, such the Hsinchu area in Taiwan and the Zhongguancun area in Beijing, China. The efforts of many city governments have been directed towards policies to identify and harness innovation that provides comparative advantage over other cities internationally. Although many have been blinded by the attraction of the Silicon Valley, technology-led, model, it is important that others have different approaches by, for example, addressing health, fashion, or the media. The issues of innovation in cities will be discussed further in Chapter 6.

Government

The debate about the role of the government in supporting innovation commonly reflects political ideology. State intervention in innovation is considered essential in many nations, including most Asian countries, but in more 'free-market' economies, such as the USA, rhetorically at least, government intervention is regarded sceptically and avoided, usually with reference to government's inability to 'pick winners'. Nonetheless, the past polarities of the views that argue, on the one hand, that interventionist innovation policies distort markets and promote inefficiencies or, on the other, are essential components of sound economic planning and effective industry policies are tending nowadays towards a pragmatic middle ground. Here it is recognized that government has an important role to play in innovation, but policies have to be selective.

Governments contribute to innovation in many ways apart from their innovation policies. A stable, growing economy enhances the preparedness of firms and individuals to invest in innovation and take risks. Effective monetary and fiscal policies are crucial in providing confidence in the future. A nation with more wealthy firms and individuals is better placed to be innovative. Good educational policies produce employees and entrepreneurs with skills to create, assess, and realize opportunities for innovation. Well-educated citizens are more capable of contributing to national debates on innovation, and determining which sciences and technologies are acceptable, and what form new products and services should take. Government investments in research – which in developed nations account on average for about one-third of total expenditures on R&D – provide many of the opportunities for innovation. These investments can often take a longer perspective than those made in the private sector. Competition policies prevent monopolies erecting barriers to innovation; trade policies increase the size of markets for innovative products and services; intellectual property laws can provide incentives to innovate; regulations in areas such as environmental protection stimulate the pursuit of innovation. Free and open access to information stored by government increases opportunities for innovation. Innovation in a highly digitally connected world is inhibited unless government acts to ensure personal privacy and encourage ethical codes of practice when it comes to the collection and use of data. Open immigration policies allow the flow of talent from overseas and are the source of diversity, which is so important for innovative thinking. Industrial relations laws can help provide equitable, secure, and participative workplaces encouraging of innovation.

Governments can encourage innovation through their procurement power: they are the primary purchaser of innovation in any nation. Public expenditures on information technology, infrastructure, pharmaceuticals, and many other areas, exceeds that of the private sector, so government purchasing is a major stimulus to innovation.

Leadership by governments can set a tone or atmosphere in which innovation is encouraged. When the political discourse is future orientated and ambitious – think of John Kennedy's plan to have man on the moon, or Harold Wilson's 'white heat' revolution of science and technology – it is more supportive of innovation than when it is relaxed and comfortable with the status quo. Public servants are more likely to support innovation when they are not fearful of censure for the slightest mistake or risk-taking behaviour.

Apart from these forms of support, many governments develop specific innovation policies. These have tended in the past, especially in scale of expenditures, to focus on R&D, usually in the form of tax credits: by spending on R&D, firms can reduce their tax bills. There have been a plethora of other types of policy designed to encourage innovation. These include demonstration schemes, highlighting the benefits of particular innovations; consultancy schemes, helping organizations to improve their ability to innovate; investment schemes offering subsidies or increasing the amount of available venture capital for innovation; and creating new intermediary organizations that help build connections between research and business.

Many justifications for government innovation policy have been championed. These include, at their most practical, fear of international competition. The US government's response to the growing dominance of Japanese competition in semiconductors in the 1980s, for example, led it to create a well-funded consortia of US manufacturers, Sematech, directed to produce competitive technologies. Many pan-European schemes in the IT industry during the same period were designed to build the capability in Europe to resist US and Japanese competition. Some policies purported to encourage innovation are simple forms of industry support – or corporate welfare, to take a less charitable perspective. Schemes around the world to continually support the ailing car manufacturing industry in marginal electoral constituencies would provide a case in point.

Much of the justification for government intervention is presented in the form of an argument about 'market failure'.

R&D, it is argued, produces knowledge that can be cheaply accessed by the competitors of those who undertake the risk of investing in it. The 'public' returns to investments thereby exceed the 'private' returns, and therefore there is a tendency towards under-investment. To address this market failure, government justifies the financial support of R&D in firms.

This form of support, which assumes the bulk of government investment in innovation policy, has several limitations. First, it is concerned with R&D, which is but one input into innovation, and in many industries and circumstances not the most important one. What is construed as 'R&D' might also be limited, and exclude important inputs to innovation such as software development and prototyping. Second, it misunderstands the investments required for public returns. The capacity of firms to access the R&D conducted by others is not costless. It requires investments to allow recipients to absorb the new ideas. Third, if market failure leads to sub-optimal investments in R&D, then there must be an optimal level, but there is little evidence about what this might be. Fourth, the delivery mechanisms of R&D support are highly generic, usually taking the form of tax credits for spending on R&D, rather than its performance. There is rarely provision for support of additional R&D to that which would be invested in without government money. The tax allowances are broadly available across industry, without the capacity to select strategic targets. Furthermore, application and compliance costs are usually resource-intensive, favouring larger, wealthier applicants rather than their generally more deserving smaller counterparts.

An additional case for government innovation policy can be made from the perspective of systems failure. Despite reservations about the dangers of the mechanical and predictable ways the national innovation systems we discuss below are viewed by government, as opposed to their commonly fluid and unpredictable reality, there is value in conceiving them from a government perspective. Government is the only actor capable of taking an overall national innovation systems view, and the only one able to influence its whole construction and function. It can assess performance, identify gaps and weaknesses, and support institutions and policies that build connections. The challenge for policy-making related to national innovation systems is that much attention is directed towards describing the components of the system, rather than what the system does, or perhaps even more importantly, what it should do.

The essential criterion for innovation policy is the extent to which it encourages and facilitates the flow of ideas across the economy and within national innovation systems and enhances the chance of their being successfully combined together and implemented. These flows of ideas occur in many, frequently unpredictable, directions: between the manufacturing, service, and resource industries; the public and private sectors; science, research, and business; and internationally within research networks or production supply chains. Innovation policy should therefore be concerned with the encouragement of the flow of ideas, the capacity of organizations to receive and use them, and the impediments to preventing effective connections between the various contributors to innovation.

Encouragement of the flow of ideas comes from open access to information and publicly funded research results, institutions that 'broker' connections between users and suppliers of knowledge, regulations that stimulate or at least fail to impede innovative investments, and judicious intellectual property laws that deal with the profound challenge of providing the confidence in their ownership to encourage trading, without providing the disincentives that occur from the award of monopoly positions. The receptivity to innovation in organizations depends upon the skills, organization, and quality of management of the recipients. Blunt policy initiatives such as R&D tax concessions are valuable only to the extent that they increase the quality and quantity of organization's capacity to select and use new ideas.

Systems

The incredible success of Japanese industry in the 1970s and 1980s led to a search for its explanation. One analysis argued that it resulted from Japan's ability to organize the various elements of its economy into a national innovation system. In this view, the Japanese government played a central role coordinating large corporations' investments in important and emerging areas of industrial technology. Japan's strength in consumer electronics, for example, was believed to have resulted from the country's highly effective Ministry of International Trade and Industry (MITI) collecting information from around the world on new technologies and organizing the efforts of large electronics firms, such as Toshiba and Matsushita, to take advantage of new opportunities. The ability of the Japanese government to do this was exaggerated, but it did play an influential role and researchers began to think about the contributions to innovation made by national institutions and characteristics and the ways they combined into a system. The search was on to try and understand the role of the major players and the most important of their interactions and provide some capacity to encourage innovation at national level.

Early research into national innovation systems took two forms. One, primarily US in focus, took an economic and legal perspective and concentrated on the nation's key institutions, including those of research, education, finance, and law. The characteristics of effective national innovation systems were seen to be high-quality research, providing new options for business; education systems that produced well-qualified graduates and technicians; the availability of capital for investments in risky projects and new and growing ventures; and strong legal protection of intellectual property. The other approach, primarily

Scandinavian in focus, was concerned more with the quality of business relationships in a society. The characteristics of effective national innovation systems were seen to be close ties between customers and suppliers of innovation, influenced by the amount of trust between people and organizations in a society and the learning this engenders.

These approaches were initially developed by academics interested in analysing and understanding the reasons why innovation occurs, and why it takes particular forms. The question emerged, for example, of why some nations, such as the USA, are particularly strong in radical innovation – explained by its strength in basic research – and why others, such as Japan, are very strong at incremental innovation – explained by efficient coordination of information exchange between customers and suppliers. The idea of national innovation systems, however, quickly took hold in government and public policy circles, as a way of prescribing and planning how institutions and their relationships could be configured. International organizations, such as the OECD, have produced numerous reports on the institutions of various nations, but these tend to be highly descriptive and static, failing to explain how national systems evolve over time. They do make the valuable observation, however, that what matters is not only the institutions that exist in a nation, but how effectively they work together.

At the same time that research on national innovation systems was blossoming, some began to ask whether the nation was the most useful level of analysis. The question was raised of why nations are often successfully innovative in some industries and regions, but not in others. The USA has Silicon Valley in California, but it also has the Rust Belt of declining heavy engineering and steel industries in the northeast. Researchers have argued the importance of regional, sectoral, and technological innovation systems. They examine the characteristics of successful regions, such as Route 128 around Boston and Cambridge in Massachusetts, Cambridge in the UK, Grenoble in France, and Daejon in Korea. They examine differences in patterns of innovation in the machine tool and textile industries. And they explore why innovation in biotechnology occurs differently from that in nanotechnology. Given the high levels of investment in innovation made by large multinational companies, operating across borders, researchers have also argued the role of global innovation systems.

The notion of systems of innovation is a helpful framework, but social systems are not engineering systems for which components and their interactions are known, planned, and constructed. Unpredictable events occur, and the systems evolve and change in unexpected ways. The early lead in biotechnology research in Harvard University, for example, was lost to Stanford University because of the election of a populist mayor in Boston who built on people's fear of unknown consequences of genetics research. What matters is thinking about the ways in which all the innovation-supporting institutions interrelate and evolve over time along with business practices and relationships. And whatever the level of analysis – global, national, regional, sectoral, technological – what is important is to understand how they relate to one another and co-evolve. The interplay between the many contributors to innovation systems is shown in the following examples of the social, cultural, political, and economic factors affecting the housing industry in Japan and research institutes in China.

Japanese housing

Japan's industrial development is rooted in a long history of deep craft traditions. These continue to pervade Japanese society, from the traditional Japanese tea ceremony and the way food is prepared, to the design of ceramics. Examining the relationship between craft skills and innovation in Japan shows the influence of social and cultural factors on innovation systems.

For centuries, all Japanese housing was produced by craftsmen using local timber. This continued until the Meiji period (1868–1912), when architectural influences and construction techniques from other countries were introduced into Japan. Japanese housing design, with its simple layout and sliding screens, also influenced Western architects such as Gropius and Corbusier.

Housing production was historically based on tens of thousands of small carpentry and building firms, each producing a few hand-crafted houses a year in conventional post and beam constructions. Tradition in housing design remains very strong and is seen in a continuing Japanese preference for the elegant and intricate timber joints that have long been a hallmark of the craftsman's art. As well as their aesthetic appeal, these joints provide rigidity in earthquakes. Yet the world's most advanced, factory-made housing industry emerged from this very conservative background. Housing innovation resulted from changing demand and new sources of supply and, while the new industry is highly automated, craft skills have been retained.

A combination of factors spurred innovation in Japanese housing following the Second World War. There were severe materials and skills shortages. There was also a large increase in demand following a wave of mass urbanization in the 1950s. Hundreds of thousands of people moved each year from agrarian communities to rapidly growing conurbations in Tokyo, Nagoya, and Osaka, stimulated by employment in new manufacturing firms and the lifestyle attractions of urban living. Mass urbanization continued throughout the 1960s and 1970s. Western styles of living became more popular, and some consumers became prepared to put their faith in large-scale manufactured products from the swiftly growing companies many of them worked for.

The drive to industrialize housing was pushed by manufacturing companies in materials and components industries, in particular steel, chemicals, plastics, and plywood. Firms turned their attention to developing new markets and several began industrial production of housing for their own workforce. Toyota opened a housing division led by the son of the company's founder. Toyota's main purpose was to create high-quality, mass-produced housing for its own workforce, and its first housing production line ran next to its car production line. In 2009, it had 6 dedicated housing factories and recently took a 50% equity stake in the country's second largest factory-made housing company – a business that has won Japan's Good Design Awards for the last 19 years consecutively.

Large industrial companies marketed homes to the new Japanese middle classes, drawing on traditions of craft in design and the benefits of quality control and reliability of industrialized production. They formed R&D centres to study technologies for housing and to assess lifestyle demands and patterns of use. Lack of space in Japanese cities focused attention as much on design and functionality in use as on the development of new materials and production processes. Yet these firms continue to offer a traditional hand-crafted tatami room even in the most contemporary of house designs, reflecting lifestyles and housing preferences that combine the convenience of modernity with the traditions of craft.

The need to rapidly produce accommodation for the 1964 Tokyo Olympic Games triggered innovation in the design and manufacture of modular bathrooms. This created an industry in which several factories each produce more than 10,000 high-quality, fully-fitted bathroom modules to individual customer specifications each month.

The housing industry invests extensively in R&D, ranging from the development of new materials, including nanotechnology coatings for building façades, to designs for multiple generations of occupants. Modular designs allow single properties to be reconfigured for young people, with space to party; young parents with close bedrooms for infants; parents with teenagers with more distant bedrooms; empty nesters, with space for guests; and the elderly, with an emphasis on ease of access.

Investment in R&D has moved from process technologies to product improvements with an emphasis on environmental and energy management. Research focuses on zero-carbon homes, safety and housing performance, and 'smart' homes with electronic sensors and controls. Firms such as Toyota have invested in the development of fuel cells and renewable sources of energy for housing. As well as the home providing electricity for the Toyota car, systems are designed so that, when required, the car can provide energy for the home. All the major producers conduct research on waste reduction and the re-use and recycling of components. Once their foundations have been laid, customized homes can be delivered, installed, and fitted out in a few weeks.

The new industry posed challenges for rural, craft-based construction by exposing its inefficiencies, high costs, and lack of innovation. Demand for craft remained high, despite most people being unable to afford a traditionally built home. Carpenters and small builders did not have the resources to invest in modern production techniques, and large industrial housing manufacturers were not interested in the fragmented rural market. The craft-based housing industry was dying and housing standards started to fall.

The problem was resolved when the forestry industry that supplied the timber for traditional craft-built homes, led by Sumitomo Forestry, took the lead in innovation. It focused effort on automating the time-consuming and expensive process of cutting traditional timber joints. Computer numerically controlled timber-cutting machines were developed and installed in some 600 micro-factories across rural Japan. Local carpenters were able to take their designs to these factories and have their timber frames produced in a fraction of the time it would have taken to cut by hand. This resulted in huge improvements in productivity, and the survival of what remained of the traditional craft industry alongside modern industrialized business.

China's science and technology institutes

The industrialization of Asia in recent decades has led to the extraordinary social and economic development of the region. Korea, for example, has been transformed from the second poorest country on Earth in the 1950s to a member of the OECD, the group of the world's 30 richest nations. Asian industrialization has required rapid developments in research, education, finance, and law to encourage the dynamic corporate and technological changes required for contemporary competitiveness. Countries such as Korea, Taiwan, and Singapore are developing coherent national innovation systems and becoming important international contributors to innovation. The models of development have varied. Korea, for example, has depended on large conglomerate companies, Taiwan on networks of small firms, Singapore on direct foreign investment by large multinationals, and China has pragmatically used all of these approaches. China is therefore a particularly valuable example of evolving innovation systems and the role of institutions in them. In East Asia, the process of development has been strongly directed by the state, and this is, of course, especially so in China.

China has experienced the most rapid and remarkable industrial development in history. From the devastation of the Second World War, civil war, and cultural revolution, it has emerged as a global manufacturing powerhouse, investing massively in science, technology, and education, and potentially challenges Western hegemony in innovation. The evolution of China's national innovation system, its features, past successes, and future challenges, can be seen in the changes affecting its science and technology research institutes. It shows the influence of political and economic factors on innovation systems and the continual challenges of change.

These institutes – which employ around 1 million people – have experienced over 20 years of major organizational reform, and substantially expanded investments over recent years. National R&D spending has increased by around 20% annually since 1999. Since China's economic reforms in the mid-1980s, these institutes have completely transformed their previous 1950s Soviet-style practice of conducting research disengaged from industry. They were encouraged in the 1980s by some severe government cuts to their budgets to focus their attention on working with business. In addition, in the 1990s a process of reforming government industrial departments led to the corporatization of around 2,000 industrial research institutes.

The reforms have been successful, but challenges still remain. The increasing nexus between research institutes and business has led to the creation of some of China's most successful start-up companies, such as Lenovo. It has redirected the focus of China's research capacity towards the market. But research institutes struggle to retain basic science investments while conducting research attractive to industry. There are complaints that commercialization of their research has distracted them from their main mission. Concern remains about the effectiveness of connections with industry. While there has been a cultural change as researchers recognize the benefits of being market-facing, establishing new forms of engagement that industry finds attractive remains elusive.

This is partly a problem of the lack of receptivity in Chinese firms. There are shortages of innovation skills in areas such as risk assessment, limiting investments in R&D and entrepreneurial start-ups. Venture capital investment tends to be concentrated in established firms and innovation-supporting investments by banks are generally made in large State-Owned Enterprises rather than entrepreneurial start-ups. Much of the focus on innovation has focused on manufacturing rather than services, and on high-technology sectors.

There is growing recognition on the part of government that innovation policy involves more than directing the research sector, and attention is being paid to improving the innovation performance of business. State direction of the former is proving much easier than the latter. Innovation systems researchers, such as Shulin Gu and Bengt-Åke Lundvall, have also questioned whether there is the extent of the social capital and trust required for the deep engagements required by researchers and businesspeople to work together and be successfully innovative.

The transformation of innovation in China experienced over the past decade has resulted from strong political leadership. It was recognized at the top levels of government that the export-led, manufacturing-based pattern of economic development behind its stunning economic growth since the 1980s would not maintain the level of growth necessary to fund China's social expectations. President Hu Jintao has called for an innovation-oriented country, pursuing a path of innovation with Chinese characteristics. The political discourse in China refers to 'harmonious growth', and the imperative for inclusive development is the most important challenge confronting innovation in China. This encompasses the need to use innovation as a means of reducing income disparities between the poor and wealthy, and the economic disparities between coastal regions and inner China. The evolution of China's national innovation system to one that allows it to compete with the West in innovation is incomplete and continuing.

Chapter 5
Thomas Edison's organizational genius

Organizations have choices on how they organize themselves for the continually evolving challenges in innovation; the structures and procedures they adopt, the staffing and incentives they use. These reflect the strategy of the organization and its innovation objectives.

Edison

Thomas Edison (1847–1931) is remembered for his inventiveness and the wide range of innovations he introduced. He held over 1,000 patents, and, amongst other remarkable achievements, he developed the phonograph, electric light bulb, and electrical power distribution, and improved the telephone, telegraph, and motion picture technology. He founded numerous companies, including General Electric. He was also responsible for pioneering a highly structured way of organizing innovation, and it is this that is our concern here.

Like Josiah Wedgwood, Edison was the youngest of a large family of modest circumstances, received little formal education, started work early, aged 12, and was afflicted by a disability, deafness, that influenced his life and work. He was similarly driven and hard-working, and shared with Wedgwood an appreciation of Thomas Paine, which also influenced his democratic worldview.

Edison could be blunt, irascible, and impatient, but he could also be personable, kind, and generous.

Edison began his working life as a telegraph operator, and started to experiment during the night shift, when he could not be observed. His first patent, an electric vote recorder, was awarded when he was 22. The notoriety of his inventions moved him from his humble beginnings into high circles. He demonstrated the phonograph to President Hayes in the White House in 1878, and was close friends with Henry Ford. He is reputed to have influenced Ford over the potential of petrol engines. His business partners included the leading capitalists of the day, such as JP Morgan and the Vanderbilts.

Edison's approach to business was relentless and ruthless. He demanded continual improvement in innovations from his employees and energetically denigrated opposition. His campaign against Alternating Current (AC) and promotion of Direct Current (DC), his preferred option for electricity transmission, sank to the unsavoury level of a publicity war over their relative merits for the electric chair. Edison did not shrink from demonstrations of electrocuting animals with AC to reveal its dangers. These included the unfortunate, if ill-tempered, Topsy the elephant, whose demise at Lunar Park was filmed by Edison for further publicity value. AC, the superior system, eventually became dominant, and the cut-throat nature of the battle between these competing technical standards clearly shows the value of owning the dominant version.

While Edison enjoyed great commercial successes, he had his fair share of failures. There were comparatively expensive and unproductive diversions into mining and the manufacture of concrete. He failed to recognize public interest in the celebrity of musicians, when for years he refused to name them on recordings. With characteristic aplomb he claimed never to have failed, but to have discovered 10,000 ways that didn't work.

Possession of intellectual property was crucial for Edison. Patents emerging from research in his laboratories were attributed to Edison irrespective of his contribution. One of his long-term assistants said: 'Edison is in reality a collective noun and refers to the work of many men.' Fiercely protective of his own patents, he occasionally disregarded the intellectual property of others. He and his business partners regularly used patents to block the development of competitors.

Fêted during his lifetime, and called a 'wizard' by the press, he faced hostile criticism from rivals. Critics included Nikola Tesla, who had every reason to be bitter. Tesla was working for Edison when he developed AC, prior to commercializing it with the Westinghouse Corporation. Tesla claimed he was not paid what he was promised. In later life, Edison regretted the way he treated him. It is speculated that the reason why Edison did not proceed with AC himself, despite having many opportunities to do so, was that he did not develop it himself; a case of the 'not-invented-here' syndrome. After Edison's death, Tesla reported for posterity his ex-boss's utter disregard of the most elemental rules of hygiene.

The way Edison organized his inventive efforts derived from his overall approach to innovation. He always pursued several lines of research, wishing to keep options open until the strongest contender emerged, when resources and effort were concentrated. By working on numerous projects simultaneously, Edison hedged his bets so future income streams did not depend upon one development. He was well aware of how when pursuing one problem it would lead to others, often completely unexpected, and understood the value of chance, serendipity and 'accident'.

He explored how ideas from different areas of research could potentially be combined and had a strategy of re-using proven components of other machines and applying them as building blocks in new designs. Edison said he readily absorbed ideas from every source, frequently starting where others left off. The development and commercialization of the light bulb, for example, combined ideas by drawing on a network of researchers, financiers, suppliers, and distributors. Although the idea of the light bulb had existed for decades, Edison, by using low-current electricity, a carbonized filament, and a high-quality vacuum, developed a product that was relatively long-lasting. His principles were to experiment and prototype as much as possible on a small scale, and to make designs as simple as possible. Once a breakthrough had occurred, he appreciated it would take a great amount of continuing research and experimentation to turn it into a successful product. He said it usually took him 5 to 7 years to perfect a thing, and some things remained unresolved after 25 years. As he said: 'genius is one percent inspiration, ninety-nine percent perspiration'.

Edison understood that most value returned to the controller of the technical system, not the producer of its individual components, who was dependent upon the system configuration. His systems thinking was most apparent in the development of the electricity distribution industry that began operating in New York in 1882. Recognizing people's apprehension of the unfamiliar, Edison cleverly mixed the new and the existing in his electricity system. He used recognizable infrastructure to deliver electricity, including putting wires underground like gas pipes and utilizing existing gas fittings in homes.

Like many of his innovations, Edison's approach to organizing his research laboratories built on the experience of others. The telegraph industry in which Edison began his career had a number of small research shops with ranges of experimental equipment. Edison had conducted experiments in one such shop in Boston, and upon his arrival in New York in 1869, he used another before setting his own laboratory in Newark to make his design of stock ticker machines.

Edison's organizational innovation lay in the range and scale of the research activities undertaken. He invested more financial and technological resources in the search for innovation than any other organization had done previously.

Edison established the Menlo Park laboratory in 1876 so he could devote himself entirely to the 'invention business'. He brought with him key workers, including a draftsman, machinist, accountant, mathematician, assayist, chemist, glass blower, and bookkeeper. Located 25 miles from Manhatten, in what was then a small village, by 1880, 75 of the 200 Menlo Park residents worked for Edison. Menlo Park began with an office, laboratory, and machine shop. Over the years, Edison added a glass house, photographer's studio, carpenter's shop, carbon production shed, blacksmiths', and additional machine shop. He also added a library.

At this time, only a few of the best universities in the USA had laboratories, and these were ill-equipped and focused mainly on teaching. Yet Edison had fine scientific equipment, including an expensive reflecting galvanometer, electrometer, and photometrical devices. Within a couple of years, his stock of tools was worth \$40,000 (\$890,000 at 2008 prices).

Edison's objective was to have all tools, machines, materials, and skills needed for invention and innovation in one place. The combination of the diverse skills in Menlo Park was assisted by their close social integration in the local community.

At its peak, Edison had more than 200 machinists, scientists, craftsmen, and labourers assisting with inventions. Work was organized into teams of 10 to 20, each working simultaneously in seeing ideas transformed into working prototypes. As everyone in the team had the same objective, communications and mutual understanding were at a premium. In 6 years at

Menlo Park, Edison registered 400 patents. He aimed for a minor invention every 10 days and a big thing every 6 months or so.

In 1886, Edison moved his main laboratory to West Orange, New Jersey, to increase the scale of his research and manufacturing capacity. West Orange was 10 times bigger than Menlo Park. Edison's biographer, Josephson, describes the reason behind the move:

> I will have the best equipment & largest Laboratory extant, and the facilities superior to any other for rapid & cheap development of an invention & working it up into Commercial shape with models patterns & special machines . . . Inventions that formally took months & cost large sums can now be done two or three days with very small expense, as I shall carry a stack of almost every conceivable material.

Edison's factory made the parts necessary for research, and research developed and made the machines for large-scale production in the factory. During the development of the phonograph over 40 years, the cylinders developed by research were first made of tinfoil, followed by a wax compound, followed by plastic. The eventual primary use of the phonograph was not the one originally imagined. As this technological and market learning occurred, the capacity to quickly scale up production of new configurations helped Edison gain substantial market share. Edison's New York factories at one stage employed over 2,000 people, one of the largest industrial concerns of the time. In contrast to the high-performance workplace laboratories, these mass-production factories operated with extensive division of labour, and the repetitive, unskilled work led to many industrial disputes.

The scale of activities in West Orange inevitably led to greater departmentalization and administration, which took up more of Edison's time. Although it was highly productive, it never matched the extraordinary output of the Menlo Park period.

Edison is quoted as saying: 'From his neck down a man is worth a couple of dollars a day, from his neck up he is worth anything that his brain can produce.' He excoriated 'pinheads' and 'lunkheads', and said: 'A man who doesn't make up his mind to cultivate the habit of thinking misses the greatest pleasure in life.' He hired graduates, but generally preferred generalists over specialists, and this is argued by some to have limited the future development of his research organization. His recruitment methods were idiosyncratic. In the early years, he would point applicants at a pile of junk and tell them to put it together and tell him when they had done so. The junk was a dynamo and those who succeeded in assembling it passed the test for employment. In later years, he compiled lengthy general knowledge questionnaires that prospective inspectors needed to pass before being promoted.

8. Edison encouraged play as well as hard work. Here workers attend a 'sing' session

Edison's style was to provide staff with a general outline of what he wanted, and then leave them to decide the best ways of achieving the objectives. He is reputed to have said: 'Hell, there are no rules here – we're trying to accomplish something.' One of Edison's staff said: 'Nothing here is private. Everyone is at liberty to see all he can, and the boss will tell him all the rest.'

He 'managed by walking about', advising and encouraging the teams. Edison worked about 18 hours a day, and the exercise he received walking from one laboratory table to another gave him 'more benefit and entertainment...than some of my friends and competitors get from playing games like golf '. Edison's biographer, Baldwin, has him 'compelled' to 'wander democratically and visibly up and down the aisles, ubiquitous, endlessly snooping, his sleeves rolled up and his unattended cigar ash dropping onto the shoulders of welders and diecutters'.

Staff worked exceptionally long hours. Tesla complained that in his first two weeks he only managed 48 hours' sleep. Legend has Edison working for five straight days and nights, but it was probably three, and it was known that the best time to contact him in the factory was after midnight. Another of his biographers, Miller, suggests: 'The capital crime of the Edison laboratory was to go to sleep. This was a source of disgrace, unless the boss could be caught napping, and then they all followed in line.' Various methods were used to dissuade the somnambulant, including the 'corpse reviver', a terrifying noise released beside the ear and the 'resurrector of the dead', which apparently involved setting sleepers alight with a small exploding substance.

It could be dangerous to work for Edison. His chief assistant, Clarence Dally, lost an arm and most of a hand during experiments with fluoroscopy during which Edison nearly lost his own eyesight. The local press reported Edison generously saying that, although Dally was unable to do any work, he would keep him on the payroll.

Josephson revealingly records the reflections of two of Edison's staff. The first, a young job applicant, was told 'Everyone applying for a job wants to know two things: how much we pay and how long we work. Well, we don't pay anything and we work all the time.' The applicant took the job. The second, a man reflecting on working for Edison for 50 years, told of his sacrifices resulting from long hours at work, including not seeing his children grow up. When asked why he did it, he responded: 'Because Edison made your work interesting. He made me feel I was making something for him. I wasn't just a workman.'

Despite these practices, which seem draconian today, Edison encouraged a creative and productive workforce. Key employees were paid bonuses from profits on inventions, although this incentive did not extend to Nikola Tesla. He socialized with staff with snacks, cigars, jokes, tales, dancing, and singing. He organized a popular midnight lunch. There was an electric toy railroad to play with and a pet bear. According to management academic Andrew Hargadon:

> The muckers [engineers] would work for days straight in pursuit of a solution, then punctuate their work with late-night breaks of pie, tobacco and bawdy songs around their giant organ that dominated one end of the laboratory.

One of Edison's assistants, quoted in Millard, said that there was 'a little community of kindred spirits, all in young manhood, enthusiastic about their work, expectant of great results', for whom work and play were indistinguishable.

Tesla complained of Edison's concentration on instinct and intuition over theory and calculation, and practices at the laboratory did occasionally seem haphazard. When searching for the best material for the light bulb filament, he experimented with unlikely materials ranging from horsehair to cork to the beards of his workers. When the breakthrough came in the carbon filament incandescent lamp, Edison's staff did not realize the extent of their discovery for several months after the event.

Nonetheless, there was focus and discipline. Edison claimed never to have perfected an invention he did not think about in terms of the service it provided, saying he found out what the world needed, then proceeded to invent. Projects had to have a practical commercial application. Famed for his 'guesswork', Edison insisted laboratory assistants kept detailed records of their experiments in over 1,000 notebooks, although this also helped with patent registration and dispute. Experimentation was extensive. 6,000 distinct species of plants, mainly bamboos, were used for carbonized filaments. 50,000 separate experiments were undertaken in developing Edison's nickel-iron battery. One of Edison's assistants, working closely with his employer, recorded 15,000 experiments into a particular problem.

West Orange possessed an extensive library of some 10,000 volumes, and Edison constantly read about biology, astronomy, mechanics, metaphysics, music, physics, and political economy. Although criticized for his disparagement of formal education, he employed two eminent mathematicians, one of whom went on to be a professor at Harvard and MIT. One of his key chemists was known as 'Basic Lawson' because of his adherence to basic scientific principles. Edison met and admired Pasteur and the German physicist and physician Helmholtz. Somewhat incongruously, George Bernard Shaw worked for Edison for a time in London.

Artefacts and drawings were important sources of creativity and communications. Edison is quoted as saying: 'Inspiration can be found in a pile of junk. Sometimes, you can put it together with a good imagination and invent something.' In 1887, his laboratory was reputed to have contained 8,000 kinds of chemical, every kind of screw, cord, wire, and needle, animals from camels to minx, feathers from peacocks and ostriches, hooves, horns, shells, and shark's teeth. Edison found it easier to think in pictures rather than words. When he was contracted by the Western Union Telegraph Company in 1877 to improve upon the telephone invented by Alexander Graham Bell, he produced over 500 sketches leading to his improved design.

As well as his internal efforts, Edison assiduously cultivated his business and research networks. He was a broker of technology, transferring research between industries. As well as his own experiments, he undertook contract research for the telegraph, electric light, railroad, and mining industries. As Hargadon puts it:

> Edison quietly blurred the line between the experiments he did for others and those he did for himself. Who was to know if a result from contract research was applied to another project or if experimental equipment built for one customer was used in work for another.

His ability to continually innovate, according to Hargadon, lay in the way he knew how to exploit the networked landscape of the time.

Edison's approach was one of trial and error, hard work, and persistence, being methodical, rigorous, and purposeful, and using prepared minds and careful monitoring. He believed innovation arose not from individual genius but collaboration, and this capacity to work together and across boundaries resulted from a supportive culture, environment, and social and industrial relations.

Edison worked on the cusp of the transition between the era of the great individual inventor and systematic, corporate

9. President Hoover called for electric lights to be switched off for a 'minute of darkness' in memory of Edison's achievements

organization of innovation. He created a form of organization for the emerging, modern technological society that was rapidly emulated by large corporations such as Bell and General Electric. In an article in the *New York Times* on 24 June 1928, it was estimated that Edison's inventions had built industries valued at \$15 billion (\$188 billion at 2008 prices). His fame was universal. President Hoover called Edison a 'benefactor of all mankind', and when he died asked people to switch off their lights for a 'minute of darkness' in his memory. His obituary in the *New York Times* on 18 October 1931 began: 'Thomas Alva Edison made the world a better place in which to live and brought comparative luxury into the life of the workingman.' An innovator can make no greater contribution.

Workplaces

As Edison so clearly showed, innovation is more likely to occur in organizations that are forward-looking, accepting of risk, and tolerant of diversity and failure. A workplace that is playful and fun and where conversation and laughter is common is more likely to be innovative than one that is highly formal, bureaucratized, and impersonal. Where expression of opinion is welcomed, ideas are not only generated more regularly, but are also implemented faster. Opposition is voiced when it has a chance to be productive rather than in the subsequent subversion of decisions.

IDEO is a company with a highly innovative workplace that emulates some of the lessons from Edison. It is a successful provider of design and innovation services, employing over 550 people in offices around the world. It has built a reputation for helping other firms to innovate in their products and services by applying creative techniques learnt in the design studio and design school environments. The company combines 'human factors' and aesthetic design with product engineering knowledge to produce products for firms from Apple to Nike to Prada. Its designs include the computer mouse, Palm Pilot, and a range of cameras and toothbrushes. It designed the whale that starred in the film *Free Willy*. IDEO has contributed to the design of over 3,000 products, and works on 60 to 80 products at any one time. IDEO has been described by *Fast Company* magazine as the 'world's most celebrated design firm'; by the *Wall Street Journal* as 'imagination's playground'; and *Fortune* described its visit to IDEO as 'a Day at Innovation U'.

For the company to deal with many diverse projects, it recruits a wide range of talent, and also enjoys special links with the Stanford University Institute of Design. It employs graduates from psychology, anthropology, and biomechanics as well as design engineering.

The leaders of IDEO have a very high profile in the international design community. They claim to have an innovative culture – 'low on hierarchy, big on communications and requiring a minimum of ego' – that uses:

> a collaborative methodology that simultaneously examines user desirability, technical feasibility, and business viability, and employs a range of techniques to visualize, evaluate, and refine opportunities for design and development, such as observation, brainstorming, rapid prototyping, and implementation.

IDEO sells its design methodologies to other companies in the form of courses and training materials. It possesses a large repository – a 'toy box' – of devices and designs from a wide range of products which staff play with when seeking solutions to new problems. It is particularly skilled at playing with creative ideas developed for one industry or project to explore their innovative application in others. Playfulness in this environment enables the cross-fertilization and the serendipitous connection and combination of unrelated ideas.

Structures

Edison pioneered a way of organizing, but organizations have wide choices in how they structure opportunities for innovation. Some choose to be highly formal and bureaucratized, others to be informal and unrestricted. Some try to do both, with parts of the organization encouraged to behave very differently from others.

One of the earliest studies of innovation in organizations, Burns and Stalker in 1961 distinguished between mechanistic and organic forms of organizing. They argued the former is appropriate to stable, predictable conditions and the latter to changing conditions and unpredictable situations. The basic general principle still applies; the way things are organized should be appropriate to the particular circumstances and aims of innovation. When technologies and markets are rapidly evolving and their future uncertain, the need – as in the case of Menlo Park – is to encourage experiment and creativity without constraining it with bureaucracy. When some of these uncertainties diminish, a more planned approach is needed to the development of projects, with highly prescribed budgets and operations geared up to deliver the innovation. Furthermore, the form of organization used often changes over time as different issues of innovation emerge. As the process of developing the innovation progresses, the supportive organizational structures move from being 'loose' to 'tight'.

R&D

R&D can be structured in very diverse ways. Many leading firms in the past relied exclusively on large corporate laboratories to undertake their research: their own large-scale Menlo Park. The archetype of this form of 'centralized' R&D was Bell Labs, which employed 25,000 at its peak and has been awarded 30,000 patents. It has received six Nobel Prizes for Physics and, amongst other things, discovered the transistor, digital switching, communication satellites, cellular mobile radio, sound motion pictures, and stereo recording. One of its basic science discoveries led to the development of radio astronomy. Founded in 1925, and based in New Jersey, it was the research group for AT&T before that company was acquired by Alcatel-Lucent. Renowned for its past strength in basic research, like many corporate laboratories, it has progressively moved towards more applied research.

The criticism of this way of organizing from a business perspective is the way research tends to be too divorced from the needs of customers and is generally too long term in orientation. In contrast, rather than having a central laboratory, other firms 'decentralize' their R&D organization structures, with laboratories located close to particular businesses or customers.

The problem with this form of structure is research tends to focus on short-term issues and miss opportunities for more radical or disruptive innovations. To try and gain the benefits of both forms, some firms combine a central laboratory with numbers of decentralized R&D laboratories, but this is a choice open only to a handful of the wealthiest.

Other organizations dispense with formal R&D organizational structures altogether. Intel, the semiconductor company, despite having a billion dollar R&D budget, has never had an internal R&D structure. It relies on networks in universities and the technology community in Silicon Valley to provide research inputs. The challenge for this 'networked' form of R&D organization is that to be receptive to knowledge from external research, organizations need to have the internal capacity to absorb it. They need skills to understand, interpret and utilize externally sourced knowledge, and often require their own in-depth expertise to attract high-quality research partners.

The organizational challenge in R&D is finding the balance between longer-term research that provides new options and insights into potentially disruptive technologies and research that deals with short-term or immediate well-defined problems. Firms often appear dissatisfied with whatever R&D structures they have. With centralized structures they feel customer needs are relegated in importance, and decentralized structures might miss potentially valuable innovations. When both forms are used, continual tensions exist over relative funding levels and ownership of projects. The problems of networked R&D lie in managing and combining inputs from multiple parties and disputes over ownership of intellectual property rights.

One strategy firms pursue to improve returns to internal R&D and access external collaborators for innovation has in recent years been described by Henry Chesbrough as 'open innovation'. The household products company, Procter and Gamble, is an example of an open innovator. It is a science-based company with a strong internal commitment to research. Its strategy is described as 'Connect and Develop' and rather than being 90% reliant on its own research investments as in the past it aims to source half its innovations from outside the company. The way it combines its own internal research with external connections is indicative of a strategy that attempts to benefit from complementary ways of organizing innovation in the same company.

The rapid growth of research capacity in China and India in recent years has the potential to change the ways many multinational companies organize their R&D. Firms create laboratories overseas to adopt their products and services to local markets, take advantage of particular local research expertise, and to create international networks of research collaboration. Many US and European firms have established substantial R&D organizations in India and China, especially in information and communications technology. The strategy these firms use can change over time. Ericsson, the Swedish telecommunications company, for example, began to invest in R&D in China in the 1980s because it helped win government contracts and was evidence of goodwill and commitment. R&D expenditure was ramped up in the early 1990s to take advantage of cheap research staff and to help adopt Ericsson products to the rapidly growing local market. Recognizing the quality and potential of Chinese researchers, both in the company and in local universities, in the late 1990s Ericsson began to locate R&D for its world markets in China. By the early 2000s, some of its R&D groups around the world were closed down and moved to China, and Ericsson's Chinese research groups became core components of the company's global R&D efforts.

New developments

R&D is one of the ways organizations create options for the future. The ways they organize their new product and service development is crucial for how successful they will be at realizing the future options they have. While R&D is generally the organizational space for scientists and technical specialists, new product and service development usually encompasses a broader range of people, including from design, marketing, and operations. These specialists help address the questions of why and how things are bought, and whether and at what cost can they be made and delivered.

There are many tools and techniques – such as 'stage-gate' systems, which operate a number of stop/go decision points in the development process – available to help plan new products and services. These are designed to help decide between competing projects and to ensure those that progress are appropriately resourced. These tools have limitations: they might be very helpful in managing the process of developing products, but they do not tell you whether they are the right products in the first place. They can also become very procedural and kill initiative.

To overcome the rigidities of bureaucracy, some organizations sanction 'bootlegging', or allowing staff to spend time working on their own projects. By giving people time – which can extend to one or two days a week – outside their formal job commitments, highly innovative companies such as Google and 3M encourage personal motivation to innovate and new ideas to emerge and blossom.

Another method used to circumvent organizational constraints to innovation is the so-called 'skunkworks'. Used initially by the Lockheed Corporation to quickly and secretly develop aircraft during the Cold War, the term is used to describe a small, tightly bound group working on a special project with considerable operational discretion within a larger organization.

Operations and production

The ways new products and services are made and delivered have been the focus of considerable innovation themselves.

Production, for example, is automated, and operations – the processes for turning inputs into outputs – have seen major innovations in the ways work is organized. Innovation in production and operations has helped create mass markets for affordable, high-quality products and services, such as automobiles, consumer goods and electronics, supermarkets, and hotel chains.

One of the key principles in organizing operations and production is Adam Smith's analysis of the division of labour. It was after reading Smith that Josiah Wedgwood saw how combining specialists in particular tasks with the new technology of steam power would improve productivity in his factory. Smith argued the division of labour is limited by the extent of the market. When markets grow large enough, benefits are gained by subdividing work and employing specialists dedicated to specific tasks rather than more expensive, broadly skilled craftsmen. He also noted specialization is a function of the division of labour, so the more work can be subdivided into discrete elements, the greater the potential to employ specialists.

Smith explained the economic benefits derived from the efficiency of the division of labour. By concentrating on a smaller range of tasks, individuals can improve their dexterity and undertake tasks more accurately and swiftly. Time is saved as there is no need to move from one task to another. When tasks are visible and discrete, machinery can more easily be devised to automate or enhance them to increase productivity.

Henry Ford used the principles of specialization and automation in developing his assembly line producing automobiles for the emerging mass market in the early 20th century. Ford's objective was tighter managerial control over production processes than previous craft forms of production allowed. His solution was the development of the mass-production line with high volumes of standardized products made from interchangeable parts. Ford learned the value of interchangeable parts from the manufacture of guns by the Colt Armory and mass production from breweries, canneries and meat packers. He combined, refined and simplified these approaches to speed up production and standardize quality on an assembly line.

His system allowed the subdivision and specialization of labour, employing unskilled or semiskilled workers on high-cost machinery dedicated to the manufacture of specific parts. Management and design were the responsibility of narrowly skilled professionals. The control of work by craftsmen was replaced by management, and the pace of work was dictated by the need to maximize use of equipment. Because machinery was so expensive, firms could not afford to allow the assembly line to grind to a halt. Buffers of extra supplies of materials and labour were added to the system to assure smooth production. Standard designs were kept in production for as long as possible because changing machinery was expensive, resulting in consumers benefiting from lower costs but at the expense of variety and choice.

Ford's friend Edison had already experienced the problem that unskilled, repetitive work caused with the industrial disputes he faced. General Motors showed Ford the limitations of his marketing approach and the benefits of producing varieties of vehicles. Alfred Sloan's approach at General Motors aimed to produce 'a car for every purse and purpose'. But the real innovation allowing both efficiency in production, wide customer choice, and better use of skills came from Japan.

After the Second World War, Toyota recognized that to realize its ambition of becoming an international carmaker it needed to harness the efficiency of American mass-production techniques and craft quality of Japanese work practices. At that time local Japanese automobile markets were small and demanded a wide variety of vehicles, production techniques were primitive in comparison to those in the USA, and investment capital was scarce. Unionized Japanese factory workers insisted they retain their skills and were unwilling to be treated as variable costs, like the interchangeable parts in Ford's and Edison's factories. Toyota understood the dangers of repetitive and boring tasks resulting in fatigue or injury to workers, with diminishing returns to efficiency.

In 1950, Toyota's President, Eiji Toyoda, spent three months at Ford's Rouge factory in the USA. He was amazed at the total output of the plant, which in one year produced over 2.5 times the number of cars made by Toyota in the previous 13 years. But while total output was impressive, Toyoda thought the system was wasteful in effort, materials and time. Toyota could not afford to produce cars with such narrowly skilled professionals and unskilled workers tending expensive, single purpose machines with their buffers of extra stocks and re-work areas. Toyoda's objectives were to simplify Toyota's production system combining some advantages of skilled craft working with those of mass production, but avoiding the high costs of craft and rigidities of factory systems. The result was the evolution of Toyota's lean production system employing teams of multi-skilled workers at all levels of the organization and highly flexible, automated machines producing large volumes of highly varied products. Rather than having buffer stocks of inventory, wasting resources, Toyota's system delivers components just-in-time to be used.

Teams of Toyota workers are given time to suggest improvements to production processes in 'quality circles': Toyota has several thousand quality circles that complete tens of thousands of small improvement projects every year. Quality circles are linked to efforts for continuous improvement (kaizen) in collaboration with industrial engineers. Emphasis on problem solving is an important part of everyone's job and on-the-job training, collective education, and self-development are all encouraged.

The success of lean production improved the whole system of designing and making cars and it made Toyota the car producer against which other manufacturing firms compared themselves. The combination of technical and organizational innovation in Toyota's production system has produced both economies of scale and scope: volume and variety.

The search for innovations that help combine economies of scale through standardization with economies of scope to satisfy diverse consumer choices is a continuing challenge. The ultimate objective in many cases is producing economically for markets of one. Toyota continues to invest in automation and new technologies such as advanced materials and techniques integrating computer-aided design with flexible computerized manufacturing systems. Unmanned supply vehicles are used to transport components and parts, and vertical computer controlled warehousing is used for storage. Despite the company's concern with skills and encouragement of quality circles, critics of the Toyota system point to the very demanding work pace required which as well as adversely affecting the health of the workforce, may inhibit innovation. The further development of its production system will depend on its accordance with what is acceptable to its employees.

Services organizations similarly look for innovation in their operations. The discount airline, easyJet, provides an example of innovative 'mass customization', or providing large-scale personalized service. The company began in 1995 with two leased aircraft and a telephone booking system. It launched a website in 1997, and by 1999, it sold its millionth ticket online. By 2005, it sold its 100 millionth. The use of the internet has been crucial to this growth and has underpinned the business model of time-dependent pricing – where prices vary with the length of time prior to booking and with demand – and highly customized customer requirements such as boarding priority, and baggage handling. It also allows the optimization of aeroplane utilization and cost reduction by not issuing tickets. It is one of Europe's largest internet retailers with 95% of its flights sold online, and also offers hotels and has a car hire partnership. All the company's documentation is stored on globally accessible servers. It has launched a desktop gadget to personalize flight information and bookings.

Another example of innovative use of operations is provided by the supermarket, Tesco, and the use of data on its 13 million regular customers. By individually classifying 25,000 products, data mining of purchasing behaviour, and the use of loyalty cards, the company creates a 'lifestyle DNA profile' of each customer. These are grouped together for particular targeted promotions. The 13 million Tesco Club Card holders are mailed four times annually with details of their rewards and vouchers on offers tailored to their profiles. Seven million variations of offers are made, and customer take up is between 10 and 25 times higher than the average 2% for direct marketing. The data is used to make sure goods available in existing and future stores are tailored to fit the profiles of local customers.

Networks and communities

Edison's development of the electrical lighting industry was an example of innovation in a technical system brought about within a network of innovators. Most innovation involves the participation of numbers of collaborating organizations and from the perspective of the individual organization this brings benefits and difficulties. The benefits lie in being able to access knowledge, skills, and other resources it does not itself possess. The difficulties lie in the absence of organizational sanction in getting others to do as you wish.

The key to effective networking is building partnerships with high degrees of trust. Trust is needed in the technical competence of collaborators, their ability to deliver what is expected of them, and their overall integrity in protecting proprietary knowledge and being prepared to admit when things go wrong. Collaborations usually start as a result of personal connections. These can break down as people move to other jobs or organizations. Effective trust between partners therefore involves the extension of inter-personal trust to inter-organization trust, with the value of the collaboration becoming institutionally engrained: legally, administratively, and culturally.

In some fields, such as open-source software, the community of users is the innovator. Here it is users of the product or service that provide new content and improvements. Despite the rhetoric of unconstrained engagement in many of these communities, a degree of organization is required. Wikipedia, for example, recognizes the efforts of its contributors to its online encyclopedia by creating a hierarchy, with significant community status accorded to wikipedians whose contributions have reached high levels of quality and quantity.

Organizations are becoming more adept at using Web 2.0 social networking sites, wikis, and blogs in their innovation activities. They are using social networking analyses by, for example, surveys or tracking email correspondence, to understand the key personal and organizational nodes in the organization and help improve decision-making. To assist communications in what are called 'massively multiparty activities' they are using virtual worlds such as Second Life, where people assume a representation of themselves as avatars. These new forms of organization pose questions about their legitimization at work – given their frequent association with 'games' – and the appropriate incentive and reward systems and skills profiles of their users.

Projects

A large part of modern economies is comprised of large, complex infrastructural projects, such as telecommunications networks, energy production and distribution, and transportation systems of airports, railways, and motorways. These projects, which commonly cost billions of dollars, entail the coordination of large numbers of firms that assemble to contribute their various skills and resources during different stages of the project's progression. They are notorious for cost over-runs and delays. The Channel Tunnel, between England and France, for example, was 80% over budget.

London Heathrow Airport's Terminal 5 (T5) was a large and highly complex project, with a budget of £4.3 billion and involving over 20,000 contracting organizations. Overseen by the British Airport Authority (BAA), the project client, airport owner and operator, it entailed the construction of major buildings, a transit system, and road, rail, and subway links, alongside the world's busiest airport working at overcapacity. T5 is the size of London's Hyde Park, and has an annual capacity of 30 million passengers. Although often remembered for the disastrous first few days of operation where British Airways misplaced 20,000 bags and cancelled 500 flights, the design and construction of the project itself was a success and was delivered to budget and on time. This success resulted from an innovative approach to managing large, complex projects.

BAA took care to learn lessons from previous projects, made sure any technologies used were already proven elsewhere, and trialled new approaches on smaller projects before applying them to T5. Use was made of digital simulation, modelling and visualization technologies to help integrate designs and construction. Underpinning the success of the T5 project was a contract between the client, BAA, and its major suppliers that differed considerably from industry norms – which were commonly adversarial – and encouraged collaboration, trust, and supplier responsibility. The risk in the project was assumed by BAA, work was conducted in integrated project teams with first-tier suppliers, and incentives were designed to reward high performing teams. Although the processes and procedures to be followed were highly specified, the project was formulated in a way that allowed managers to confront the unforeseen problems that inevitably arise in complex projects flexibly and on the basis of their previous experience.

The lessons from T5 are that success in large, complex projects involves standardized, repetitive and carefully prepared routines, processes, and technologies and the capacity to be innovative to be able to deal with unexpected events and problems. Organizing projects involves a judicious balance between performing routines and promoting innovation.

Creative people and teams

As Edison showed in Menlo Park, innovation involves a team effort, bringing together different ideas and expertise. The construction of teams involves decisions about the most appropriate balance of skills given the problems faced. It also entails deciding on the comparative value of organizational memory – keeping people in teams together – and refreshment – bringing in new skills. Teams that work together for long periods tend to become introspective and become immune to innovative ideas from outside. Teams that are newly constructed or contain many new members have to learn to work together effectively and develop a modus operandi. Harmony in teams holds many virtues, but sometimes it is important for innovation to have disruptive elements – the grit in the oyster – asking difficult questions and shaking things up.

The structure of teams has to reflect their objectives. Those devoted to more radical innovation need more creativity and flexibility in objectives, with the freedom to respond to emerging and potentially unforeseen opportunities. They often need heavyweight support from higher levels in the organization as their objectives do not quickly add to the bottom line and as a result they are vulnerable to criticism and cost-saving exercises. A balance has to be found between incentives for individuals and teams. The factors that encourage innovation team effectiveness are often subjective – to do with professional satisfaction and recognition. Those that inhibit performance are more instrumental – to do with project objectives and resource limitations. As Edison found out, employees will work extraordinarily hard when given the incentive of interesting, rewarding, and appreciated jobs.

Creativity is not only important to design companies such as IDEO. Innovation in all organizations relies on creative people and teams to produce new ideas, and creativity is an issue that infuses the whole of the world of work. By its stimulation of innovation, many contemporary organizations see the encouragement of creativity as core to their development and competitiveness. Creativity provides a means of making work more attractive, improving the engagement and commitment of existing staff, and a winning strategy in the 'war for talent' amongst highly skilled and mobile employees.

Creativity has an individual and a group component. Psychologists tell us about the characteristics of creative people, and how imaginative ideas emerge from individuals with the capacity to think differently and see connections and possibilities. Creative individuals are said to have a tolerance for ambiguity, contradiction, and complexity. Cognitive scientists, such as Margaret Boden, argue that creativity is something that can be learnt by everyone and is based in ordinary abilities that we all share, and in practiced expertise to which we can all aspire.

Organizations expend large amounts of time and resources on creativity training, and constructing incentives and rewards for individual creativity. They are also concerned with the promotion of creativity in groups and with formulating the most conducive team structures and organizational processes and practices. Groups bring together the disparate perspectives and knowledge that are valuable for creativity and are essential for the new combinations in innovation. Recent research into creativity has focused more on the organizational and business circumstances that encourage creativity and the systems and strategies that shape its manifestation.

Creative ideas become useful innovations when they are successfully applied. Creativity in itself may be inspiring, stimulating, and beautiful, but it has no value economically until it is manifested as an innovation. It takes different forms in incremental and radical innovations. Incremental innovations usually involve a form of creativity that is more structured, managed, and deliberate. Radical innovation requires creativity that may be unbounded by existing practices and ways of doing things.

People

Leaders

Innovation rarely occurs in organizations without the commitment and visible support of their leaders, although those leaders may have little idea of the specific nature of new developments. One of the key aspects of leadership is encouragement for the creation of new ideas and their implementation. Leaders find resources for support and offer protection from the opponents of innovation. When new ideas threaten the status quo, established interests will inevitably oppose them. As Machiavelli says in *The Prince*:

> There is nothing more difficult to plan, nor more dangerous to manage than the creation of a new order of things ... Whenever his enemies have the ability to attack the innovator they do so with the passion of partisans, while the others defend him sluggishly, so that the innovator and his party alike are vulnerable.

One of the lessons of renowned leaders of innovative organizations, such as Edison, is they create a supportive culture where staff are encouraged to try new things and are not discouraged when they fail. In 1948, the chairman of 3M, William McKnight, summarized his approach that characterized the company's strategy for decades to follow...

> As our business grows, it becomes increasingly necessary to delegate responsibility and encourage men and women to exercise their initiative. This requires considerable tolerance. Those men and women to whom we delegate authority and responsibility, if they are good people, are going to want to do their jobs in their own way.
>
> Mistakes will be made...Management that is destructively critical when mistakes are made kills initiative. And it's essential that we have many people with initiative if we are to continue to grow...

A nervous young manager who had led a failed project once offered Henry Ford a resignation letter. Ford's response was that he was not about to let someone go and work for a competitor after learning a valuable lesson with his money.

Managers

As well as supportive leadership at the top of organizations, particular innovations need enthusiastic and powerful managerial 'champions' or sponsors with significant decision-making responsibility. As well as being good at managing teams, coordinating technical/design issues, and implementing processes and decisions, innovation managers also have to be skilful at advocating the virtues of the innovation, lobbying for its support, and creating a vision of what it will do and contribute.

Boundary spanners

One of the most important individual roles in innovation is the boundary spanner, the person capable of communicating and building bridges between and within organizations. In manufacturing firms, this person used to be known as the technological gatekeeper. These people are avaricious acquirers of information – attained through reading and attending conferences and trade shows – and skilled at communicating useful information to the part of the organization that needs it. Organizations sometimes find it difficult to justify appointing boundary spanners. Their remit to travel, go to conferences, and talk to lots of people is sometimes unappreciated by those bound to a desk or workbench. But their role is highly beneficial to innovation.

Everyone

One of 3M's most successful innovations was the Post-It Note. Due recognition has been made of the developers of the technical core of the innovation – its non-sticky glue. And sufficient opprobrium has been dealt to the company's marketing department that claimed no one would buy it. But too little credit has been accorded to the people in the organization that recognized the product's potential and encouraged its development. Following the rejection of the idea of the Post-It from the marketing department, the product's developers sent samples to the secretaries of the company's General Managers. The secretaries immediately saw the value of the product and elicited the support of their bosses in getting the idea developed.

Innovation affects everyone in an organization and to a greater or lesser extent is everyone's responsibility. The computerization of many traditional craft skills, such as toolmaking, provided opportunities for de-skilling or re-skilling jobs. Many employers took the de-skilling path – as in the case of numerical control machine tools– but subsequently learned the advantages of re-skilling and giving the shop-floor worker discretion over the tasks they perform. This reflects the capacity of people to change and productively and creatively respond to innovation, if given the chance. It gives them, in Edison's words, the pleasure of cultivating the capacity to think. The potential of innovation derived from the factory floor has led some to describe them as laboratories and places for experiment.

An important tool used to encourage innovation is the use of reward and recognition programmes. Many organizations have suggestion schemes, and companies such as IBM and Toyota elicit hundreds of thousands of ideas from employees. These can be rewarded financially or by peer recognition. Often the most effective form of recognition is the implementation of the idea by the organization. The capacity of individuals throughout the organization to have innovative ideas and pursue their implementation shows that innovation leadership is not only the responsibility of those with high hierarchical positions.

Innovators in all forms are best supported in organizations whose commitment to human resource development and training attracts, rewards and retains talented managers and staff unafraid of change, and placates those who are fearful of it. Innovative organizations have the appointment procedures, payment and incentive systems, and career progression paths, to ensure appropriate staffing for innovation. While some people thrive on creating innovation, and need to be encouraged and rewarded, others are better at developing the procedures for its application, requiring different forms of recognition. Others still are temperamentally fearful of innovation, or at least too much change, seeing it as threatening and can suffer stress and poor performance as a result. A reputation as an innovative organization is very attractive to potential recruits who want to be innovative, and selection mechanisms should vet unsuitable appointments. Employees who find it disturbing need to be supported and guided through the introduction of innovation.

Technology

During the 1960s, the research of Joan Woodward into factory organization in the southeast of England began to explain the relationship between technology and organization. She showed how organization varied according to core underlying technology, whether production took the form of small or large batches, mass production, or continuous flow processes. The view that organization results from the technology used – technological determinism – has been discounted by research showing the extent to which choices can be made, a view to which Joan Woodward subscribed. Nonetheless, technology is highly influential and there is a relationship between the ways industries are organized and the extent to which they can benefit from innovation through the division of labour. The products and services of industries vary considerably and production and operations techniques vary accordingly.

Innovation technologies

Edison knew the value of high-quality scientific instrumentation, on the one hand, and 'junk', odd pieces of machines, and a vast array of unusual materials, on the other. These machines and artefacts stimulate innovation. Just as Edison's many sketches aided his thinking and improved the communication of his ideas with others, the creation of tangible designs and prototypes focuses efforts and builds connections between people with different skills and perspectives. In many cases, ideas for innovation grow organically and iteratively around emerging and increasingly focused designs.

Information and communications technologies move design and connections across boundaries into a digital world where Edison's aim for 'rapid and cheap development of an invention & working it up into Commercial shape', occurs in ways he could not imagine.

Digital technologies combine design and manufacture in Computer-Aided Design/Computer-Aided Manufacturing systems. Digital design information on new products is transferred to the equipment used to make them. The designs are guided by the system as to what is possible to manufacture. The internet, local area networks and enterprise resource planning systems help organizations combine different inputs from people with very different skills.

Developments of massive computing power, software that allows the merger of different data sets, and new visualization technologies used extensively in the computer games industry, has led to a new kind of technology supporting innovation. 'Innovation technology' (IvT) is so called because it helps combine various components of the innovation process. It is being used to improve the speed and efficiency of innovation by pulling together different inputs within and between organizations. IvT includes: virtual reality suites used to help customers design new products and services; simulation and modelling tools used to substantially improve the speed of new designs; E-science, or Grid computing, building new communities of scientists and researchers, and helping them to manage collaborative projects; sophisticated data-mining technology used to help understand customers and manage suppliers; and virtual and rapid prototyping technology used to improve the speed of innovation. Together, these technologies are being used to make customers on the one hand, and scientific researchers on the other, combine more effectively in decisions about innovation.

By moving experiments and prototyping into the digital world, IvT allows firms to experiment cheaply and 'fail often and early'. IvT is also very important in the design of large, complex systems, such as utilities, airport infrastructures, communications systems, where it is not usually feasible to test full-scale prototypes.

One of the most important aspects of IvT is how it assists the representation and visualization of knowledge and its communication across different domains, disciplines, professions and 'communities of practice'. By way of illustration, compare the design of a new building using traditional methods and IvT.

The use of IvT makes complex data, information, perspectives, and preferences from diverse groups visible and comprehensible. Virtual representation assists architects to visualize their eventual designs and help to clarify clients' expectations by giving them a good understanding of what a building will look and feel like before work begins. Clients can 'walk through' their virtual building getting a sense of its layout and 'feel' prior to a brick being laid. IvT informs contractors and builders of specifications and requirements, and allows regulators, such as fire inspectors, to confidently assess whether buildings are likely to meet regulatory requirements. IvT can allow various players in the innovation process, suppliers and users, contractors and subcontractors, systems integrators and component producers, to collaborate more effectively in the delivery of new products and services.

Using IvT can produce some quite dramatic innovations. A great many died in the World Trade Center in 2001 because occupants trying to get down the fire escape stairs were trapped

10. Engineering and design increasingly use computerized visualization and virtual reality tools

with firefighters going up them. New ways of getting people out of tall buildings in extreme events were considered for the replacement Freedom Tower in New York. Computer simulations and visualizations of the behaviour of buildings and people in emergencies led fire engineers to believe that the safest form of egress was by means of the lift. Changing entrenched views on safety to one where the message is: 'In case of fire, use the lift' requires a great deal of persuasion of building owners and occupants, engineers and architects, firefighters and fire regulators, and insurers. Mutual, shared comprehension of this radical change was assisted by moving detail from complex drawings and data sets to readily understandable computerized images. Fire engineers have used a range of new simulation and visualization technologies to help transform these diverse parties' understanding of safety in tall buildings and encourage the exploration of innovative approaches to rapid evacuation.

Chapter 6
Building a smarter planet?

We began this book with an illustration of innovation at the beginning of the Industrial Revolution. We end it with a speculative glimpse at what the future may hold. The challenges and opportunities for innovation are immense. As well as creating new sources of wealth from ideas, innovation is essential if we are to cope with climate change, provide better water and food, improve health and education, and produce energy sustainably. It will be essential for our continued co-existence on an increasingly crowded planet.

The innovation processes that will be used have become increasingly more complex. They have evolved from the activities of 18th-century entrepreneurs such as Josiah Wedgwood, to the formal organization of research in the 19th century and large corporate R&D departments of the mid- and late 20th century, to the present-day involvement of multiple contributors in distributed networks of innovators supported by new technologies.

The keys to future innovativeness will lie in an organization's ability to foster creativity and to make decisions and choices on the basis of being well prepared, informed and connected. The many sources of ideas – employees, entrepreneurs, R&D, customers, suppliers, and universities – will continually produce opportunities to innovate. The challenge lies in encouraging, selecting and configuring the best ideas from them. To explore how organizations in the future might respond to these challenges we turn to the example of IBM, a company that is using a distributed innovation process, and is preparing the ground for its further development in the future. We choose IBM because, in contrast to organizations of a more recent vintage, such as Microsoft and Toyota, or smaller, entrepreneurial firms, such as IDEO, IBM has had a long history of transforming itself. IBM has continually faced challenges, some of its own making, others imposed, and its survival in the future as much as the past has depended upon innovation. IBM has shown how it can change its innovation processes as it adapts to and shapes demands for new products, services, and technologies, and responds to emerging challenges. Although we cannot predict whether it will succeed in the future, the company shows us some of the aspects of a contemporary networked innovator; in the ways it supports and encourages expertise, internal and external connections, and decision-making in uncertain situations.

Future thinking: the case of IBM

In 2006, IBM conducted a 'massively parallel conference' or 'Innovation Jam', which involved creating a web portal and inviting its staff to post ideas about four carefully identified future fields of potential development. The results were remarkable: in two three-day phases, over 40,000 suggestions were made by 150,000 IBM employees, family members, business partners, clients and university researchers from 104 countries. The interactive 'Jam' process saw ideas being discussed and refined, rated and scored as they were reduced in number to 36 and eventually 12. Nick Donofrio, IBM's head of innovation and technology at the time, says the process saw ideas getting momentum, evolving and turning into something completely different as those involved took over and changed them. Over $70 million was allocated in funding 10 new businesses as a result of the 2006 Jam, generating around $300 million in revenues in two years. In this way, IBM uses the internet to draw on the creativity of a massive community of potential innovators. It also uses a web portal, ThinkPlace, where employees can identify, share, and be rewarded for innovative suggestions. By these means, IBM has developed a way of systematically attracting ideas for, and making choices about, innovation on a huge scale.

Many of these proposals derive from the company's 200,000 scientists and engineers. To encourage creative science and engineering, and to ensure strong external connections amongst its technological leaders, IBM has created the positions of Distinguished Engineers and IBM Fellows. Around 650 staff currently hold these high-status positions within the organization. To become a Distinguished Engineer, individuals need to demonstrate a sustained record for invention and innovation, and recognition amongst internal and external peers. The position of IBM Fellow is the highest accolade for technical achievement in the company, and Fellows are given considerable latitude to pursue their own areas of research. Appointments to these prestigious positions are important motivators for career development and signals of achievement.

As part of its efforts to continually build its technological expertise, IBM has an Academy of Technology, founded in 1989, modelled on the US National Academies of Science and Engineering. Its purpose is to advise IBM's executives on technical trends, directions, and issues, and to develop and connect the IBM technical community across the world. The Academy produces reports, has an annual conference, and plays an important role in raising awareness about emerging trends and building and sharing knowledge.

The company's search for innovation extends well beyond its boundaries and it considers itself part of an 'innovation ecology' with a myriad of external relationships and connections. This ecology includes independent software providers, technical standards bodies, universities, government agencies, and customers. A regular series of publications – Innovation Outlooks – is produced to demonstrate thought leadership and help engage with communities that contribute to IBM's development and use of innovation.

IBM filed 4,186 new US patents in 2008, more than any other company. At the same time, as an indicator of the extent of its efforts to engage with the innovation community, it announced plans to increase by 50%, to more than 3,000 annually, the publication of its inventions and technical contributions, making this research freely available. By being 'open' in its innovation, and providing intellectual property for use by others, it demonstrates its expertise and helps build the scale of its technologies and markets as other parties develop complementary new products and services.

As an example of the way IBM engages with its customers, it has a system of Client Technology Advisers (CTAs) who develop long-term relationships with key clients, offering advice and developing roles as trusted guides to strategic developments in their industries. IBM benefits from this arrangement by helping shape major new investments at their early stages.

An example of IBM's attempts to lead in the development of new ideas is seen in the concept of 'services science'. This emerged in 2004 as a description of the shift towards complex services and systems in many industrial sectors and markets. The concept was adopted by a number of IBM's clients and collaborators such as BAE Systems and HP. IBM also works with the academic community, sponsoring research projects and university symposia aimed at exploring what it hopes will become a new discipline. IBM's encouragement of courses in Services Science, Management and Engineering (SSME), has seen their adoption worldwide in around 400 universities.

One method IBM uses to choose amongst its many potential opportunities for innovation is its Emerging Business Opportunities (EBO) process. EBO was created in 2000 to improve the company's ability to explore new technologies, such as virtual worlds, and quickly respond to new business opportunities. EBOs are usually managed in decentralized businesses, comprising small teams to encourage focus, and are expected to quickly generate financial returns, demonstrating there is a market for its products and services. Some EBOs, such as cloud computing – which provides immense computing power over the internet on a needs basis – are such major initiatives they are incubated centrally in the organization. The greatest incentive for encouraging new ideas is often the reward of seeing them put into practice, and the EBO process, which enjoys top-level management support, signals IBM's intent to systematically pursue the best of them.

IBM's organizational efforts to build and support its innovation process are complemented by its approach to using technology to build connections and make decisions. This is one of the key features of the company's Smart Planet strategy.

IBM's Smart Planet strategy

Launched in 2008, this strategy recognizes that dealing with complex and emerging problems – in areas such as energy, health and the environment – requires understanding of the relationships within and between systems. This depends on the ability to monitor performance and make sense of massive amounts of data.

The strategy is based in part on the potential for using data from large numbers of sensing and monitoring instruments, including mobile devices. An indicator of the ubiquity of these instruments is IBM's estimate that there are around 1 billion transistors for every human on the planet. An example of this form of instrumentation is radio frequency identification (RFID) devices. RFID offers innovative approaches to managing supply chains and logistical systems, for example by tracking meat from farm to shop, helping to ensure fresh food is readily available for consumption.

IBM's strategy is to connect these sensors, to allow systems and objects to communicate together, creating what is known as the 'internet of things'. To create intelligent ways of finding and delivering solutions to problems using the Internet, decisions about how to design, configure, and operate systems will need to be made, using the powerful analytical capabilities of supercomputers and cloud computing. This allows data to be mined and patterns to be recognized, such as when insurance companies spot patterns in millions of claims, or police make correlations in forensic evidence to identify patterns of crime. Their analyses and diagnoses can lead to new understanding of the ways systems perform and evolve, and better management of resources. The use of innovation technologies, such as simulation techniques and the visualization of their results, can also offer the promise of decision-makers engaging with a wide range of stakeholders when making choices.

Examples of the potential value of the Smart Planet approach can be found in energy, transportation, and health. The scale of the challenge in energy provision is seen in IBM's estimate that the current inability to manage and balance electricity supply and demand in the US results in enough electricity being lost annually to provide the power for India, Germany and Canada. By utilizing an extraordinary density of instruments for measurement and monitoring, innovation can optimize energy supply and demand. The innovation opportunity is to develop new systems in which everything can be instrumented and analysed in real time: from the meter in the home, through the distribution network, to power plants. These 'smart grids' have the potential to allow better decisions and make energy supply more efficient, reliable, and adaptive to changing needs.

Similar issues are seen in the case of transport systems. According to IBM, congestion on US roads costs around $80 billion annually, with over 4 billion lost working hours and wastage of nearly 3 billion gallons of fuel, which produces huge carbon dioxide emissions. Innovation in public policy using instrumentation of traffic systems has resulted in improvements in many cities. Traffic congestion charges in Milan, for example, have a scale depending upon the level of pollution emitted by individual vehicles. When a vehicle enters the city, cameras almost instantaneously signal to a database identifying the model and relevant charge band. Stockholm's smart traffic system, which uses cameras and lasers to identify and charge vehicles according to the time of day, has reduced congestion by nearly 25% and emissions by about 12%. The complexity of the problems being confronted is seen in the way IBM worked with 300 different organizations in developing the Stockholm system.

Another example is seen in a means for addressing the notorious traffic problems in Chinese cities. Siemens, the German electronics and engineering company, is working with Chinese researchers to optimize traffic flow based on positional data provided automatically by drivers' mobile telephones.

An example of the Smart Planet approach in healthcare is provided by IBM's project with Google Health and Continua Health Alliance aiming to create telemedicine systems that allow individuals and families to track health information by streaming data from medical devices. RFID is also used to confirm the authenticity of medical supplies, reducing errors and improving compliance with medical regulations and procedures. It is being used by Siemens to track the number of sterile pads used in surgical operations to prevent any being left behind in the patient, and to monitor the temperatures of blood supplies throughout the process of donation, cell concentration, storage and use.

Many of the challenges of energy, transportation, and health will be addressed in cities. In 1900, 13% of the world's population lived in cities. By 2007, the majority were urban dwellers, and this is expected to increase to 70% by 2050, by which time the world's population will have risen from 6 to 9 billion. The difficulties for city planners and authorities are seen in the way cities are responsible for 75% of global energy consumption and 80% of greenhouse gas emissions.

An indication of the way IBM intends to develop services that address problems in urban systems is seen in the example of the creation of its Business Analytics Research Centre, announced by Sam Palmisano – IBM's President, Chairman and CEO – at the company's Smarter Cities Summit in Berlin in 2009. The German Centre will employ 100 scientists to research urban systems. The Centre's international, interdisciplinary, and inter-organizational research is an example of the new distributed models needed to tackle pressing contemporary problems.

IBM describes itself as a globally integrated enterprise, with global scale in many core activities, but whose delivery of services focuses on local market requirements. The company has the advantage of drawing on the huge amount of its science and technology expertise found around the world. At the same time, rather than concentrating R&D in the USA or a few major global hubs, the choice of Berlin for the new Centre shows how IBM aims to distribute its expertise closer to markets, especially in this case those of Eastern Europe.

The Centre therefore operates as a hub in a distributed research network, collaborating with more than 300 German mathematicians, consultants and software specialists. It links with IBM's global research centres, drawing on the broad expertise that exists on the complex interactions within and between systems. The Centre's work is collaborative, builds on IBM's services science approach, and uses innovation technologies to create and analyse new mathematical models and simulate the behaviour of the systems of systems that are cities. To understand energy requirements, for example, it is insufficient to model the increased use of electric cars without understanding its implications for electricity supply.

The new innovation model shows that no single organization on its own – even one the size of IBM – has the depth and breadth of capability to tackle the problems facing urban systems. It recognizes how solutions to the problems of population growth, health, energy, and transportation are interconnected and global in reach.

How successful the Smart Planet strategy will be in helping solve these problems will depend on the resolution of significant technical, organizational, social and political questions. Simple, robust, fail-safe technological systems are needed. New skills are required to analyse the mountains of data, interpret patterns, and produce insights to assist in making near real-time decisions. New forms of participation have to be developed to engage people in developing and implementing innovations.

IBM's Smart Planet is the evolving strategy of one organization, whose success or failure has yet to be enacted. IBM, like all organizations, has made mistakes in the past and will do so again in the future, and we have yet to see whether and how the strategy will work. The concept of a smarter planet, however, is an enticing one when considering the broader role of innovation in the future, and IBM gives us some clues as to what directions we might move in. The company shows us how the interconnections encouraging innovation are based on talented and knowledgeable people, deeply engaged within innovation ecologies, guided by imaginative strategies, and facilitated by technology. A smarter planet would use resources more efficiently and effectively. It would use new organizational approaches and technologies in highly connected institutions, innovation processes, and ways of working, and be better positioned to address current and emerging challenges.

Smarter institutions

Governments

As well as pursuing the innovation policies discussed in Chapter 4, encouraging the stock and flow of innovation, national governments need high levels of inter-governmental coordination: internationally, regionally, and locally.

The use of innovation to deal with many contemporary problems requires more resources and skills than can be mustered by individual nations. Some challenges, such as controlling greenhouse emissions, simply cannot have autonomous solutions and have to be addressed in international forums. Balancing national self-interest with the need for international approaches will pose an increasing innovation policy challenge. Furthermore, as social well-being and economic prosperity become driven ever more by the productive use of creativity and knowledge, there are profound implications for relationships and disparities between nations. Existing inequalities may become accentuated as the technologically, institutionally and organizationally rich nations pull further away from those not so advantaged. Inter-governmental institutions have to monitor and develop policies to address any such problems.

Many important decisions about innovation are made not at the national level but by increasingly powerful municipal authorities and regional governments that vigorously compete with each other domestically and internationally to attract investment and talent. Expertise in domestic inter-governmental coordination and collaboration is also essential to effective innovation policy.

The privatization in many countries of previously publicly held assets in energy, transportation, and telecommunications has removed a direct lever governments once possessed to improve innovation. Instead, new regulatory authorities have been established and their roles in supporting innovation in the private sector have to be explored and extended. Government's role is complicated by the ways boundaries of the public and private realms have become blurred with the creation of public/private partnerships. There are mutual benefits to be achieved from this form of organization, including access to resources for investment in innovation that may otherwise be unavailable. But the ownership and control of innovation assets and knowledge may be shaped by different incentives that may result in tensions between private gain and public good. Government innovation policy has to be formulated on the basis of deep engagement with business, and understanding of the strengths and shortcomings of the contributions it can make.

There are huge future opportunities for innovation in government services. These include, for example, telemedicine, using computers and the internet to assist medical diagnoses at home. Telemedicine is used in Australia to deliver health services to remote communities. In the UK, it is used to provide monitoring of elderly patients preventing them having to be in hospital. In India, mobile equipment is driven to impoverished villages where diagnoses are made via electronic links to urban hospitals, providing a level of health care to which the rural poor previously had no access.

As part of their contributions to innovation, governments have opportunities to use the new technologies that offer the means for more inclusiveness and participation in decision-making by citizens in designing and delivering the services we demand. By creating proposed new health centres in virtual worlds prior to their actual construction, for example, inputs can be elicited from health professionals and patients to produce better designs.

One of the most crucial areas of policy-making lies in the processes by which choices are made by governments about where to focus innovation investment to sustain future prosperity. No nation has the resources to innovate in all areas and trade-offs are needed between competing demands for scarce resources. Governments have to establish sophisticated approaches for making choices, whilst ensuring that enough is invested in a broad spread of areas to keep options open and allow nations to absorb useful ideas developed elsewhere. Decisions about what and what not to prioritize have to involve extensive discussions with business, social and environmental groups, and informed public debate in efforts to build consensus about the future.

The importance of innovation for government, and the difficulties in building the necessary connections and making good choices, requires broad and deep innovation policy-making skills. These extend comprehension of the importance and nature of innovation throughout the apparatus of government, and help develop a 'whole of government' approach. Greater appreciation of the contributions and difficulties of innovation will help address the very high aversion to risk in the public service. In recognition of its more widespread, distributed, and inclusive nature, public policy requires better forms of measurement of innovation – moving on from the partial and often misleading indicators of R&D expenditure and patenting performance – and new approaches and skills are needed in this area. Tools, such as social network analysis can be used, for example, to measure changing patterns of connectivity. Innovation policy-making has to recognize that innovation is a continuing challenge with no simple 'solutions'. As it evolves new issues arise and policies need to change in response.

Universities

To contribute more effectively to innovation, universities have to be better at encouraging knowledge exchange and the internal and external flow of ideas. They should move beyond a restricted model of technology transfer in the form of formal intellectual property protection, licensing and start-up companies, and welcome the many opportunities collaboration provides for the creation and transfer of new educational and research services. Their strategies will find multiple ways of engaging with stakeholders in business, government, and the community, and yet still continue to be driven by scholarly values. They will educate and employ people able to work in multiple ways in research, business, and government, and build connections between different parts of innovation systems, encouraged by the mobility of broadly skilled graduates and enhanced through the use of eScience.

Few universities have the resources to provide universal service across all the academic disciplines, and most benefit from making choices and broadly specializing. Some may strengthen their positions by having a local flavour, while others see their role as nodes in global research and education efforts. Concentration in particular areas provides the depth of expertise attractive to the best collaborators in research and business, and partnerships can fill gaps where institutions have decided not to participate.

Universities have a continuing role to play in producing the large-scale research tools and instruments for science and engineering to encourage discovery, allow people to scout in unknown areas and see and measure the things that others cannot: perhaps increasingly in the design and delivery of services. They provide leadership in formulating common standards needed by innovators to help launch new products and services in dynamic industries.

The provision of 'rehearsal space' and collaborative laboratories for deep and sustained conversations and engagement in ideas generation and testing with business, government, and the community is one of the most important innovation supporting roles for universities. Researchers will continue to work with all the academic rigour and independence of their discipline, but through these conversations many will become comfortable as members of distributed teams exploring inter-disciplinary interfaces and the social and economic consequences of their work. Highly accomplished at providing the physical and organizational structures and incentives for scholarly recognition and career progression, universities will need to explore better spaces and methods for encouraging and rewarding such engagement.

Business

When economies and technologies are rapidly changing and volatile, the value of firms' capacity to accept and implement radical and disruptive ideas increases. In such circumstances,

'This really is an innovative approach, but I'm afraid we can't consider it. It's never been done before.'

11. Some challenges of innovation may always be with us

the best strategies are those that are experimental and dynamic and achieve a judicious balance between exploiting existing ideas and exploring new ones. These strategies rely on continual investments in human capital and in research and technology.

Innovation, in the words of Lou Gerstner, previous CEO of IBM, needs to be engrained in the DNA of the organization. As seen in the case of IBM, the performance of exceptional innovators and teams should be rewarded, but the responsibilities and opportunities for innovation are everyone's.

Continuing investments in R&D and the absorptive capacity it produces remain crucial, as does the ability to trade and broker knowledge within innovation ecologies. Corporate connections to sources of new ideas require long-term partnerships with universities around the world, deep embedding within innovative cities and regions, and effective management of supportive innovation technologies.

The breadth of these ecologies is extended as traditional distinctions between industries become blurred as knowledge, insights, and skills across sectors are transferred and combined to produce novel offerings. Much value creation in the manufacturing industry, for example, lies in design services. The services sectors and universities are collaborating in innovative ways, such as in IBM's SSME initiative. Innovation is a condition for success in the creative industries – such as new digital media, entertainment, and publishing – whose content is critical, for example, for innovative product and services companies involved in mobile telephony. The resource industries, such as agriculture and mining, depend on innovation to improve efficiencies and assist product improvements, and whose innovations in water management, for example, have broader applications.

The ideas for innovation in business will come from diverse and often unexpected sources in new and unforeseen combinations.

The effect of the global financial crisis of 2008/2009 on connections between innovators and sources of capital will take many years to be fully understood. In the short term, innovation investments will undoubtedly be adversely affected, and in the long-term trust between the financial and productive sectors will need to be rebuilt. New forms of governance of risk will be needed to oversee greater ethical and responsible decision-making and improve risk management of complex innovations.

Small and medium-sized organizations may increasingly be the progenitors of breakthrough technologies, using their advantages of speed, flexibility, and focus over larger ones. Compared to large, publicly traded companies, small firms can take and bear unusual risks. And not being so constrained by the organizational rigidities of large firms, they can more easily develop and trial novel business models and processes. Small and medium-sized organizations will combine their behavioural advantages with the greater resources found in large firms in new forms of innovation network and collaborative partnership. Larger organizations will continually experiment with attempts to emulate the entrepreneurial environments of smaller units, as seen in IBM's EBO process.

As Edison knew, innovation has to be organized in ways appropriate to its objectives. The benefits from the unbounded search for ideas, where chance and serendipity can produce so many rewards, have to be balanced with organizational focus and direction. There are far more opportunities than can be afforded, and choices need to be made that shape and direct the skills organizations use and resources they invest. Skills in the strategic management of innovation which help them

make those choices will become amongst the most prized
by business.

Smarter innovation

As in the time of Wedgwood, innovation will result from the
combinations of ideas, but these ideas are ever more widespread
and distributed around the globe, and their integration can
increasingly be assisted by the use of technology. As Wedgwood
understood so well, innovation combines 'supply side'
considerations – that is, sources of innovation such as research and
technological developments – and deep appreciation of market
demand. Smart innovators are immersed in understanding the
changing patterns and meaning of consumption, and the values
and norms that underlie decisions to purchase innovative
products and services. These patterns are affected by
globalization and are fluid in nature. A generation raised on
conspicuous consumption, whatever its real costs, may be despised
by another concerned for sustainability. Recognizing the
capacities of new technologies to include increasing numbers of
contributors to innovation, including communities of users,
greater appreciation is needed of their motivations and how
their energy and insights can be most effectively used.

Innovation strategy in companies of all sizes and sectors has to
move beyond the planned, sequential models of the industrial age,
and the forms of corporate R&D laboratories that produced
Stephanie Kwolek's discovery. It has to account for opportunities
that arise in unexpected places, high levels of uncertainty and
great complexity, where organizational learning through
collaboration is the key to survival and growth. The limited
financial measurements and accounts used by business in the
past – such as return on capital and quarterly reports to
shareholders – have to be supplemented with indicators more
meaningful to innovation and organizational resilience. What,
for example, is the value of the options for the future that

organizations possess through doing research? What innovations being explored and developed have the potential to account for major parts of the organization in 10 to 20 years? How has an organization's capacity to learn been improved through research investments? What is the value of being a trusted collaborator, and ethical employer and sustainable producer?

Economic thinking benefits from evolutionary approaches that see risk, uncertainty, and failure in innovation as normal and move us away from linear and planned to open, emergent, and highly connected systems. The value of ideas and learning become recognized as the most important drivers of economic growth and productivity. The importance of the exploration of new inter-disciplinary combinations between science, arts, engineering, social sciences and humanities, and business is appreciated and the need for mechanisms and skills for building connections across organizational, professional, and disciplinary boundaries is emphasized. Attention is placed on improving the connections and performance of innovation systems and ecologies. These ecologies may form unimagined new combinations: anthropology may inform local energy production and distribution; philosophy may influence semiconductor circuit design; the study of music may affect the provision of financial services.

Innovation technologies intensify innovation. The instrumentation of trillions of devices and sensors embedded in the physical world contribute to the unimaginable amounts of data available to be used by the new technologies of design in the virtual world to create and improve the products and services we want and enhance the experiences we desire.

Innovation has to deliver non-damaging or environment-enhancing products and processes. Innovation and sustainable development will need to become two sides of the same coin. Many sustainability challenges – climate change, water resources management, genetically modified agriculture, waste disposal, marine ecosystem protection, and biodiversity loss – are persistent and have no complete solution, lacking a clear set of alternatives and little room for trial and error. They are characterized by contradictory certitudes among protagonists and strategies for dealing with them involve coping rather than solving and searching for what is feasible not optimal. Lessons from the study of innovation can be applied to deal with these persistent problems, including the facilitation, structure and management of cooperation and connectedness, the management of risk and assessment of options, and the use of collaboration tools such as social networking technologies. Furthermore, the use of innovation technologies can help model and simulate the implications of decisions, and their visualization capacities help communications and the informed involvement of diverse parties to assist participative decision-making.

Smarter individuals

How will we deal personally with the way innovation is changing? Whether we work in the private or public sector, community groups, or as a member of the public, how can we be smarter in the way we develop and use innovation? Increased technological literacy will certainly improve our effectiveness in the massively connected world. But we shall also have to become more adept at encouraging creativity, dealing with change, communicating across boundaries, and putting ideas into practice. Intuition and judgement, tolerance and responsibility, diversity of interests and cross-cultural sensitivities are needed. Our capacities to think about new ideas, play with them by tinkering, testing, and prototyping, and rehearsing, implementing or doing them, have to be balanced. Our scepticism and critical faculties should be attuned to questioning: 'this is the way it is', and capacities accentuated for articulating: 'this is the way we want it'. We shall demand the rewards that Edison's laboratory workers experienced – although perhaps without the exhausting hours or fear of the corpse reviver. Indeed, with the wealth created through our knowledge, we expect job satisfaction in enriching workplaces, conducive to diversity, which fit in with our lifestyles and family circumstances and choices.

We should ensure the inventors and innovators that contribute so much – the Stephanie Kwolek's of the world – are recognized in a way that sport stars and entertainers are appreciated today.

Innovation is a restless process that brings with it continual uncertainty about its success and failure. It can be threatening as well as rewarding. How well we respond to it depends upon how open-minded and cooperative we are, how prepared we are to accept risk and give space for the unusual and work with others that are differently minded. This will be influenced by the culture of organizations and the quality of leaders that recognize that job security and toleration of failure are crucial to innovation, that no one holds all the answers, progress is collaborative, and reputation lies in modesty in claims and professionalism in delivery.

The outcomes of innovation are not always beneficial, and their consequences often cannot be foretold. Adding lead to petrol solved the problem of engine knocking, but left a disastrous environmental legacy. Thalidomide reduced morning sickness in pregnant mothers, but induced disabilities in their babies. The dangerous divorce between action and consequence was clearly seen in the global financial crisis of 2008/9 where financial innovations were introduced without any checks and balances or consideration for their implications. Concern for the implications of innovations have to be paramount amongst those who seek to introduce them.

The massive amounts of data available on individuals, to other people, corporations, and the state, also increase the responsibilities of those designing and managing innovation. Innovation – in information use and other areas such as genetics – requires deep ethical considerations and highly visible and accountable practices, and alert and responsive regulations. Simulation, modelling, and virtualization technologies provide huge opportunities for improving innovation processes, but their responsible use depends on the skills and judgement of people immersed in the theory and craft of their professions and trades. Innovation requires people to be informed, vigilant and responsible employees, customers, suppliers, collaborators, team members, and citizens. Andrew Grove, the founder of Intel, said that in our uncertain world only the paranoid survive, but it will be the discerning and informed, not the mistrusting and fearful, that will see us through. Immanuel Kant said that science is organized knowledge; wisdom is organized life. The future of innovation – where its benefits flow and costs are curtailed – lies in the wise organization of knowledge.

References

W. Abernathy and J. Utterback, 'Patterns of Industrial Innovation', *Technology Review*, 80(7) (1978): 40–7.

N. Baldwin, *Edison: Inventing the Century* (New York: Hyperion Books, 1995).

W. Baumol, *The Free-Market Innovation Machine: Analyzing the Growth Miracle of Capitalism* (Princeton, NJ: Princeton University Press, 2002).

C. Bilton, *Management and Creativity: From Creative Industries to Creative Management* (Oxford: Blackwell, 2007).

M. Boden, *The Creative Mind: Myths and Mechanisms*, 2nd edn. (London: Routledge, 2004).

T. Burns and G. Stalker, *The Management of Innovation* (London: Tavistock Publications, 1961).

H. Chesbrough, *Open Innovation: The New Imperative for Creating and Profiting from Technology* (Cambridge, MA.: Harvard Business School Press, 2003).

C. M. Christensen, *The Innovator's Dilemma: When New Technologies Cause Great Firms to Fail* (Boston, MA.: Harvard Business School Press, 1997).

A. Davies, D. Gann, and T. Douglas, 'Innovation in Megaprojects: Systems Integration in Heathrow Terminal 5', *California Management Review*, 51(2) (2009): 101–25.

M. Dodgson, D. Gann, and A. Salter, 'The Role of Technology in the Shift Towards Open Innovation: The Case of Procter & Gamble', *R&D Management*, 36(3) (2006): 333–46.

M. Dodgson, D. Gann, and A. Salter, ' "In Case of Fire, Please Use the Elevator": Simulation Technology and Organization in Fire Engineering', *Organization Science*, 18(5) (2007): 849–64.

M. Dodgson and L. Xue, 'Innovation in China', *Innovation: Management, Policy and Practice*, 11(1) (2009): 2–6.

G. Fairtlough, *Creative Compartments: A Design for Future Organisation* (London: Adamantine Press, 1994).

C. Freeman, *The Economics of Industrial Innovation*, 1st edn. (London: Pinter, 1974).

C. Freeman, *Technology Policy and Economic Performance: Lessons from Japan* (London: Pinter, 1987).

C. Freeman and C. Perez, 'Structural Crises of Adjustment: Business Cycles and Investment Behaviour', in G. Dosi, C. Freeman, R. Nelson, G. Silverberg, and L. Soete (eds.), *Technical Change and Economic Theory* (London: Pinter, 1988).

D. Gann and M. Dodgson, *Innovation Technology: How New Technologies Are Changing the Way We Innovate* (London: National Endowment for Science, Technology and the Arts, 2007).

L. Gerstner, *Who Says Elephants Can't Dance: Inside IBM's Historic Turnaround* (New York: Harper Business, 2002).

S. Gu and B.-A. Lundvall, 'China's Innovation System and the Move Toward Harmonious Growth and Endogneous Innovation', *Innovation: Management, Policy and Practice*, 8(1–2) (2006): 1–26.

A. B. Hargadon, *How Breakthroughs Happen: The Surprising Truth about How Companies Innovate* (Cambridge, MA.: Harvard Business School Press, 2003).

C. Helfat, S. Finkelstein, W. Mitchell, M. Peteraf, H. Singh, D. Teece, and S. Winter, *Dynamic Capabilities: Understanding Strategic Change in Organizations* (Malden, MA: Blackwell, 2007).

R. Henderson and K. B. Clark, 'Architectual Innovation: The Reconfiguration of Existing Product Technologies and the Failure of Established Firms', *Administrative Science Quarterly*, 35(1) (1990): 9–30.

M. Josephson, *Edison* (London: Eyre and Spottiswoode, 1961).

C. Kerr, *The Uses of the University* (Cambridge, MA.: Harvard University Press, 1963).

R. K. Lester, *The Productive Edge* (New York: W. W. Norton & Co., 1998).

B. A. Lundvall (ed.), *National Innovation Systems: Towards a Theory of Innovation and Interactive Learning* (London: Pinter, 1992).

F. Malerba, *Sectoral Systems of Innovation: Concepts, Issues and Analyses of Six Major Sectors in Europe* (Cambridge: Cambridge University Press, 2004).

K. Marx, *Capital*, Vol. 1 (Harmondsworth: Pelican, 1981).

A. Millard, *Edison and the Business of Innovation* (Baltimore, MD: Johns Hopkins University Press, 1990).

F. Miller, *Thomas A. Edison: The Authentic Life Story of the World's Greatest Inventor* (London: Stanley Paul, 1932).

R. Nelson and S. Winter, *An Evolutionary Theory of Economic Change* (Cambridge, MA.: Belknap Press, 1982).

R. Nelson (ed.), *National Innovation Systems: A Comparative Analysis* (New York: Oxford University Press, 1993).

D. F. Noble, *Forces of Production: A Social History of Industrial Automation* (New York: Oxford University Press, 1986).

C. Paine, *Who Killed the Electric Car?*, documentary film, Papercut Films (2006).

S. Palmisano, 'The Globally Integrated Enterprise', *Foreign Affairs*, 85(3) (2006): 127–36.

J. Quinn, 'Interview with Stephanie Kwolek', American Heritage. com, 18(3) (2003).

E. M. Rogers, *Diffusion of Innovations*, 4th edn. (New York: The Free Press, 1995).

R. Rothwell, C. Freeman, A. Horley, V. Jervis, Z. Robertson, and J. Townsend, 'SAPPHO Updated – Project SAPPHO, Phase II', *Research Policy*, (3) (1974): 258–91.

Royal Society, *Hidden Wealth: The Contribution of Science to Service Innovation* (London: Royal Society, 2009).

K. Sabbagh, *Twenty-First-Century Jet: The Making and Marketing of the Boeing 777* (New York: Scribner, 1996).

J. A. Schumpeter, *The Theory of Economic Development: An Inquiry into Profits, Capital, Credit, Interest and the Business Cycle* (Cambridge, MA.: Harvard University Press, 1934).

J. A. Schumpeter, *Capitalism, Socialism and Democracy* (London: George Allen & Unwin, 1942).

S. Smiles, *Josiah Wedgwood: His Personal History* (London: Read Books, 1894).

A. Smith, *An Inquiry into the Nature and Causes of the Wealth of Nations* (London: Ward, Lock and Tyler, 1812).

D. Stokes, *Pasteur's Quadrant: Basic Science and Technological Innovation* (Washington, DC: Brookings Institution Press, 1997).

D. J. Teece, 'Profiting from Technological Innovation: Implications for Integration, Collaboration, Licensing and Public Policy', *Research Policy*, 15(6) (1986): 285–305.

J. Uglow, *The Lunar Men: Five Friends Whose Curiosity Changed the World* (New York: Farrar, Straus and Giroux, 2002).

J. M. Utterback, *Mastering the Dynamics of Innovation: How Companies Can Seize Opportunities in the Face of Technological Change* (Boston, MA.: Harvard Business School Press, 1994).

R. Williams, *Retooling: A Historian Confronts Technological Change* (Cambridge, MA.: MIT Press, 2002).

J. Womack, D. Jones, and D. Roos, *The Machine that Changed the World: The Story of Lean Production* (New York: Harper, 1991).

J. Woodward, *Industrial Organization: Theory and Practice* (London: Oxford University Press, 1965).

Further reading

On Josiah Wedgwood:

B. Dolan, *Wedgwood: The First Tycoon* (New York: Penguin, 2004).

On Joseph Schumpeter:

T. McGraw, *Prophet of Innovation: Joseph Schumpeter and Creative Destruction* (Cambridge, MA.: Harvard University Press, 2007).

On the innovation process, and the ways it is organized, managed, and changing:

M. Dodgson, D. Gann, and A. Salter, *Think, Play, Do: Technology, Innovation and Organization* (Oxford: Oxford University Press, 2005).

M. Dodgson, D. Gann, and A. Salter, *The Management of Technological Innovation: Strategy and Practice* (Oxford: Oxford University Press, 2008).

On the economics of innovation:

J. Fagerberg, D. Mowery, and R. Nelson (eds.), *The Oxford Handbook of Innovation* (Oxford: Oxford University Press, 2006).

On the history of innovation:

N. Rosenberg, *Inside the Black Box: Technology and Economics* (Cambridge: Cambridge University Press, 1982).

D. Edgerton, *Shock of the Old: Technology and Global History since 1900* (London: Profile Books, 2006).

On innovation strategies:

M. Schilling, *Strategic Management of Technological Innovation* (New York: McGraw-Hill/Irwin, 2005).

On entrepreneurship:

G. George and A. Bock, *Inventing Entrepreneurs: Technology Innovators and their Entrepreneurial Journey* (London: Prentice Hall, 2009).

M. Wright, B. Clarysse, P. Mustar, and A. Lockett, *Academic Entrepreneurship in Europe* (Cheltenham: Edward Elgar, 2007).

Data on international R&D and innovation performance:

National Science Foundation; Science and Engineering Statistics: http://www.nsf.gov/statistics

Organization for Economic Cooperation and Development; Science, Technology and Patents, Statistics Portal: http://www.oecd.org

Index

Index

Innovation

Visit the
VERY SHORT
INTRODUCTIONS
Web site

www.oup.co.uk/vsi

➤ **Information** about all published titles

➤ News of **forthcoming books**

➤ **Extracts** from the books, including titles
not yet published

➤ **Reviews** and views

➤ **Links** to other **web sites** and main
OUP web page

➤ Information about **VSIs in translation**

➤ **Contact** the editors

➤ **Order** other **VSIs** on-line

ARCHAEOLOGY
A Very Short Introduction
Paul Bahn

This entertaining Very Short Introduction reflects the enduring popularity of archaeology – a subject which appeals as a pastime, career, and academic discipline, encompasses the whole globe, and surveys 2.5 million years. From deserts to jungles, from deep caves to mountain tops, from pebble tools to satellite photographs, from excavation to abstract theory, archaeology interacts with nearly every other discipline in its attempts to reconstruct the past.

'very lively indeed and remarkably perceptive … a quite brilliant and level-headed look at the curious world of archaeology'

Barry Cunliffe, University of Oxford

'It is often said that well-written books are rare in archaeology, but this is a model of good writing for a general audience. The book is full of jokes, but its serious message – that archaeology can be a rich and fascinating subject – it gets across with more panache than any other book I know.'

Simon Denison, editor of *British Archaeology*

www.oup.co.uk/vsi/archaeology

SOCRATES
A Very Short Introduction
C. C. W. Taylor

Socrates has a unique position in the history of philosophy. It is no exaggeration to say that had it not been for his influence on Plato, the whole development of western philosophy might have been unimaginably different. Yet Socrates wrote nothing himself, and our knowledge of him is derived primarily from the engaging and infuriating figure who appears in Plato's dialogues. In this book, Christopher Taylor explores the relationship between the historical Socrates and the Platonic character, and examines the enduring image of Socrates as the ideal exemplar of the philosophic life; a thinker whose moral and intellectual integrity permeated every detail of his life, even in the face of betrayal and execution by his fellow Athenians.

www.oup.co.uk/isbn/0-19-285412-7

THE RUSSIAN REVOLUTION

A Very Short Introduction

S. A. Smith

The Bolshevik's aim was nothing less than to destroy an entire social system, and to replace it with one superior to any hitherto seen in human history. In this Very Short Introduction to the Russian Revolution, Steve Smith explores the far-reaching reverberations of the drive to modernize a bakward, isolated Russia. Using the shift in perspective brought about by the fall of Communism and the passing of the twentieth century, he reflects on such fundamental issues as the role of ideology in the Revolution, the interplay of emancipatory and enslaving elements in Bolshevik practice, and the influence of Russian culture on the development of Soviet society.

www.oup.co.uk/isbn/0-19-285395-3

ART HISTORY
A Very Short Introduction
Dana Arnold

This clear and concise new introduction examines all the major debates and issues using a wide range of well-known examples. It discusses the challenge of using verbal and written language to analyse a visual form. Dana Arnold also examines the many different ways of writing about art, and the changing boundaries of the subject of art history. This book explains how the traditional emphasis on periods and styles originates in western art production and the problems that can arise from this.

A wide variety of topics are explored such as the canon of Art History, the role of the gallery, 'blockbuster' exhibitions, the emergence of social histories of art (Feminist Art History or Queer Art History, for example), the impact of photography, and the development of Art History using artefacts such as the altarpiece, the portrait, or pornography, to explore social and cultural issues such as taste, religion, and politics.

http://www.oup.co.uk/isbn/0–19–280181–3

EVOLUTION
A Very Short Introduction
Brian Charlesworth and Deborah
Charlesworth

In less than 500 years the relentless application of the
scientific method of inference from experiment and
observation, without reference to religious or
governmental authority has completely transformed our
view of our origins and relation to the universe.

This book is about the crucial role of evolutionary biology
in transforming our view of human origins and relation to
the universe, and the impact of this idea on traditional
philosophy and religion. The purpose of this book is to
introduce the general reader to some of the most
important basic findings, concepts, and procedures of
evolutionary biology, as it has developed since the first
publications of Darwin and Wallace on the subject, over
140 years ago. Evolution provides a unifying set of
principals for the whole of biology; it also illuminates the
relation of human beings to the universe and each other.

http://www.oup.co.uk/isbn/0-19-280251-8